Einstein

AND OUR WORLD

CONTROL OF NATURE

Series Editors

Morton L. Schagrin
State University of New York at Fredonia

Michael Ruse
Florida State University

Robert Hollinger
Iowa State University

Einstein

AND OUR WORLD

SECOND EDITION

DAVID C. CASSIDY

Humanity Books

an imprint of Prometheus Books
59 John Glenn Drive, Amherst, New York 14228-2197

Published 2004 by Humanity Books, an imprint of Prometheus Books

Inquiries should be addressed to
Humanity Books
59 John Glenn Drive
Amherst, New York 14228–2197
VOICE: 716–691–0133, ext. 207
FAX: 716–564–2711

08 07 06 05 04   5 4 3 2 1

Library of Congress Cataloging-in-Publication Data

Cassidy, David C., 1945–
    Einstein and our world / David Cassidy. — 2nd ed.
        p. cm. — (Control of nature)
    Includes bibliographical references and index.
    ISBN 1–59102–256–8 (pbk. : alk. paper)
        1. Einstein, Albert, 1879–1955—Influence. 2. Physicists—Intellectual life.
3. Relativity (Physics) 4. Quantum theory. 5. Physicists—Biography. I. Title.
II. Series.

QC16.E5C37 2004
303.48'3—dc22

2004007648

Printed in the United States of America on acid-free paper

CONTENTS

LIST
OF
ILLUSTRATIONS

Illustrations

Table

ACKNOWLEDGMENTS

I would like to thank Joseph Haberer, Spencer Weart, Margaret C. Jacob, Esther Sparberg, Jeremy Bernstein, and my students at Hofstra University for their helpful comments and suggestions. I would also like to thank Ann O'Hear and the staff of Humanity Books for their fine work.

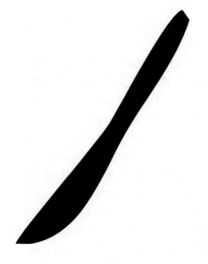

THE
EINSTEIN
AGE

Napoleon and other great men of his type: they are makers of empires. But there is an order of men who get beyond that. They are not makers of empires but makers of universes. . . . Ptolemy made a universe which lasted 1400 years. Newton, also, made a universe which lasted 300 years. Einstein has made a universe, and I can't tell you how long that will last.

—George Bernard Shaw

Who was this maker of universes, and what has been his influence on our world?

Born in Germany during the early imperial era in 1879, Albert Einstein died seventy-six years later in Princeton, New Jersey. His death in 1955 came one decade after the defeat of Nazi Germany in World War II and the dropping of the first atomic bombs on Japan. As a physicist, essayist, and man of moral principle, Einstein has left a profound and far-reaching impact upon his world and ours—so much so, in fact, that Einstein's name, like that of Sir Isaac Newton before him, has been closely associated with a twentieth-century transformation as profound as that in Newton's day from the Renaissance to the Age of Enlightenment.

Newton, in his work in physics and on methods of research, built upon the contributions of his predecessors during a century of scientific genius. His enunciation of the laws of motion, force, and universal gravitation enabled for the first time a complete comprehension of the motions of the moon and planets and promised a complete scientific understanding of the everyday physical world. Such striking success of human reason encouraged a period of profound faith in human capability and renewed optimism for unlimited progress in all fields of human endeavor—political, economic, social, and cultural. Fully enlightened by the laws of nature and reason, it was argued, humankind could not fail to achieve perfection of the individual and happiness and equality for all.[1] How different Einstein's and our worlds are from those of Newton and the eighteenth century! Einstein's most profound contributions to science occurred in two of the greatest intellectual contributions of the twentieth century: relativity theory and quantum mechanics. These are sciences that come into play when we encounter events at scales far beyond the everyday realm of Newtonian science: enormously small sizes, those of atoms, nuclei, and elementary particles; enormous velocities, those close to the speed of light; enormous distances, those on the cosmic scale; and enormous masses, those of the sun and planets, black holes, and neutron stars. Here strange things occur, contrary to common sense and to Newtonian physics: moving lengths contract, moving clocks slow down, moving masses increase, and space curves in the presence of matter. Here, too, what we can know about nature and how we know it are no longer as we would expect from everyday experience. There may even be limits to what we can know, a circumstance unknown to Enlightenment science.

By the same token, the unbridled faith in human progress and reason spawned by Newtonian science has died a painful death. The darker side of human nature, uncovered by Sigmund Freud and modern psychology, has found ample manifestation behind the barbed wire of state terror unleashed by Hitler, Stalin, and their successors. The innocent optimism of the Enlightenment, shaken in the trenches of World War I, succumbed to the horrors of World War II, ushering in an era of cold-war fear and cynicism. Even as the modern era and the cold war reach their "postmodern" end, fear and cynicism remain, terrorism—the antithesis of human reason and dignity—is on the rise, the evils of ethnic persecution continue, and faith in the liberating benefits of scientific and technological progress remains in doubt.

If science is the process by which people attempt to gain a rational understanding of the natural world around them, then both scientists and nonscientists alike often interpret scientific results as general truths about their world and their place within it. Because of this, Einstein the man, living amid tremendous social transformation and upheaval, became for many the embodiment of, and spokesman on behalf of, human dignity and reason. Yet, for many others, Einstein the physicist symbolized the incomprehensible new world in which they now found themselves. To a public that suddenly felt itself adrift, cut off from the ties of traditional beliefs, mores, and meanings, Einstein's theory of relativity seemed to indicate that nature itself supported this state of affairs in the form of "relativism." There are no absolute truths, Einstein was interpreted as saying; all knowledge and morals are relative and therefore subjective and ephemeral. An ultimate understanding of the meaning of human existence and destiny is impossible. Unlike the faith in human reason displayed by some of their Enlightenment counterparts, many twentieth-century thinkers and cultural figures saw in Einstein's work confirmation of a world moving into an uncertain and meaningless future, and this was reflected in their own works, discussed later in this book.

For others, the success of Einstein's physics, which lies at the very foundations of our understanding of nature, resulted in an association of his name with the triumph of twentieth-century science—both its enormous success and the enormous economic, political, and military power that derived from it. When *Time* magazine named Albert Einstein its "Person of the Century" in 1999, it declared, "He was the preeminent scientist in a century dominated by science. The touchstones of the era—the Bomb, the big bang, quantum physics, and electronics—all bear his imprint."[2] Einstein's discoveries and theories and those of his successors have not only satisfied our basic need to know and have inspired others to similar heights in other fields of endeavor, but they have also found practical applications in nearly every area of daily existence—from the new technologies of microchip semiconductors, lasers, photoelectric cells, and nuclear energy to the astounding breakthroughs in cosmology and the origins and fate of the universe. Yet the very success and power of Einstein's science have also brought renewed fear, anxiety, and antiscientific reaction, both within and far beyond scientific circles.

How should *we*, living a half-century after Einstein's death and years after the end of the nuclear cold war that occupied his last days, comprehend this man and his influence on our world? The answer requires a closer look at the interconnections between history, people, and science.

EINSTEIN'S INFLUENCE

Do great people make history, or does history make great people? This question has been asked many times and answered in numerous ways over the past two centuries. Whereas the influential nineteenth-century thinkers G. F. Hegel and Karl Marx provided perhaps the most affirmative answers, respectively, to each side of the question in their time, most historical research has since transcended the narrow perspective of great persons distinct from the historical circumstances in which they live and work. Until recently, however, the same could not be said for the history of science, especially for the history of a science such as physics.

Physics was often considered a "hard science," not because of its supposed difficulty, but because it seemed impenetrable by "external factors": historical, economic, ethical, and cultural circumstances.

Physicists themselves perceived their work as a search for universal laws of nature, a search that was driven solely by "internal" criteria, rules, and methods of research that seemed fundamentally ahistorical. Following the early physicists' lead, historians regarded the development of physics as a self-propelled triumph of human ingenuity, a linear progression from ignorance and doubt to truth and certainty achieved by a succession of heroic geniuses who, like a caste of priestly saints, contemplated the truths of nature alone and in isolation from worldly cares. Great people acting alone, it seemed until recently, do make great science history.

Since the end of World War II, but especially since the work of Thomas S. Kuhn, many historians of science have become increasingly unable to view the history of a science such as physics in total isolation from the broader social and cultural contexts in which it exists.[3] As a result, the history of science has achieved a greater sophistication and depth as well as a new professional status as a legitimate branch of historical research. Adapting historical methods to its subject, the new profession treats science like all other human endeavors, as a cultural phenomenon, and scientists like other historical figures, as human players in a fascinating historical drama.

The transformation has not occurred without its difficulties and controversies. Debates have raged over such matters as the relative importance of "internal" and "external" factors, how exactly they are related, and if the division into "factors" is even appropriate; the role of sociological, irrational, and personal elements in the development of science; and the very nature of scientific theories. But the new perspective has also freed the science historian to explore new uses of institutional and intellectual history; the development of new forms of nonheroic biography; new approaches engendered by feminist and nontraditional perspectives on Western research; and an appreciation of the significance of science education, research schools, and social, ideological, and moral commitments.

Einstein has still to benefit fully from the new history of science. While historians have explored the details of his private love life and the origins, growth, and acceptance of his science, they have directed far less attention to the other side of the cultural interaction: to the impact of his ideas and discoveries outside his field and on the broader culture and times.[4] While realizing that history may help make great people, most science historians have yet to appreciate that great people and their work help make great cultural history.

That we are justified in submitting both the origins and the impact of Einstein's work to historical scrutiny is again suggested by a figure of comparable stature, Sir Isaac Newton. Like Einstein, Newton worked essentially alone and entertained little personal need for other people, and the intricacies of his work were and are fully accessible only to the technically initiated; yet the results of that research were fundamental and far-ranging throughout and far beyond physical science. Moreover, both saw their successes not so much as theirs alone but as the culmination and extension of the work of others who had gone before them. If he was able to see farther than his predecessors, Newton once remarked, then it was because he stood on the shoulders of giants.

Like a shock wave expanding from an exploding epicenter, the impact of Einstein's work on our world spread first throughout physics and the physics profession. It formed a new foundation for understanding nature that then spread to other physical sciences and, in so doing, transformed the research enterprise. Einstein's early work also coincided with and helped to propel development of the new profession of theoretical physics and its incorporation, along with experimental research, into the powerful discipline we know today as contemporary physics.

Through both the new science and the new discipline of physics, Einstein and his colleagues exerted a profound impact upon the broader culture and society. But the impact contains an irony. Borrowing from physicist Hermann Helmholtz, historians Christa Jungnickel and Russell McCormmach have titled their two-volume study of the rise of theoretical physics *The Intellectual Mastery of Nature*.[5] Mastery, they suggest, has enabled human control over nature. The irony is that—in view of the potential misuse of such breakthroughs as nuclear energy and genetic engineering, as well as new social dislocations caused by recent technologies—some people now wonder whether in fact it is nature (or, more precisely, knowledge thereof) that controls us!

Although much remains to be learned, the emphasis in this book is on the impact of the Einstein the man and the physicist upon the world in which we live, rather than upon the origins and technical nature of the new physics—relativity theory and quantum mechanics—to which Einstein especially contributed. The impact is viewed as having occurred in two ways: (1) through new ideas about nature and public understanding or misunderstanding of them and (2) through the rise of the theoretical physicist to professional status and social recognition, owing again to a great extent to Einstein and his work. Chapter 2 briefly summarizes Einstein's early life and the historical and intellectual contexts of Einstein's *annus mirabilis*, or "miraculous year," of 1905. Chapter 3 is a very brief, nontechnical summary of those "miracles": special relativity, quantum theory, and the atomic hypothesis. Chapters 4 and 5 examine the reception of the new physics within physical science, the technologies it spawned, and the rise of the new profession to prominence by mid-century. Chapter 6 considers the uses and abuses of the relativity and quantum theories by a public in turmoil. Chapter 7 explores cultural influences occurring in twentieth-century art, literature, and philosophy. Finally, Chapter 8 returns to Einstein the individual in an age of dictatorship, genocide, and weapons of mass destruction.

This short monograph can offer only a brief overview of the results of recent research on these very complex issues. It is intended for nonspecialists. As indicated, this story concerns the nature and impact of the intellectual transformation wrought by Einstein and his physics. How this transformation arose within the technical details of Einstein's physics can be only briefly summarized here. Several excellent, nontechnical presentations of Einstein's theories are listed in the bibliography. Readers desiring fuller accounts of the physics, the history, or any other topics treated here are also referred to the bibliography.

DÉBUT
DE
SIÈCLE

T he four decades that separated Einstein's birth in 1879 from the end of World War I encompassed both the climax and the demise of the old European order, as well as the turn of the twentieth century. In that period, and in the process becoming an important driving force behind it, Einstein completed his formulation of relativity theory and made his most fundamental contributions to quantum physics. In so doing he rose from obscurity to the pinnacle of his profession as a theoretical physicist in Berlin.

Einstein's work not only hastened the rise of theoretical physics to public prominence, but it also provided a striking new comprehension of fundamental natural processes. The decades between the two world wars encompassed an age in desperate search for a new understanding of itself and the world around it and for a new worldview to replace the previous one, which had so recently and abruptly faded in war and social unrest. It was in the context of that era that Einstein's early work exerted its greatest and most lasting impact.

HEYDAY OF THE OLD ORDER

The years following the unification of Germany in 1871 witnessed unprecedented growth in German and European industrial power, along with extraordinary militaristic nationalism and imperialistic expansion on all sides, leading up to World War I.

Yet it was also an age that looked to the future with enormous optimism. Not unlike some aspects of the digital computer revolution a century later, the newly invented electric light, telephone, phonograph, automobile, and airplane, along with the newly discovered x-rays, radioactivity, and radio waves, captured the imaginations of all Europeans and encouraged a renewed faith in material progress through technological advance. As industrialization and new science-based industries, such as the sprawling German chemical and electrical concerns, spread prosperity to ever-wider sectors of society, they also encouraged expectations of greater freedom and democracy. Rulers could no longer deny a political voice to industrial workers, while local bureaucrats could no longer deny women admission to university degree programs (but universal suffrage still remained years away).

Yet the dark clouds of the twentieth century were also gathering. Darwin's theory of evolution challenged the cozy, traditional notions of humanity's origins and place in nature and inspired callous adherence to "social Darwinism." Radical writers such as Friedrich Nietzsche and Fyodor M. Dostoyevsky argued the meaninglessness of human affairs, while on the southern fringe of Europe old tribal rivalries and hatreds were bursting into murderous violence.

As ethnic wars flared to the south, the anti-Semitism of the Dreyfus Affair in France spread across Europe during the 1890s, even entering the platforms of established political parties. In the same period Sigmund Freud uncovered the significance of irrational urges in the human psyche—and all this in an age enamored of rational science, technological progress, and petty Victorian propriety! The juxtaposition is remarkable, yet characteristic of the century to come. It is no accident that both Einstein and Freud received their greatest public attention during the chaotic years following World War I. Within the realm of fate it is perhaps also no accident that Freud published his earliest fundamental work, *The Interpretation of Dreams*, in the same year, 1900, that Einstein graduated from the Swiss Polytechnic, at the dawn of the twentieth century.

Three aspects of Einstein's work and approach at the turn of the century aligned with broader movements and longings of the period to render both Einstein and his work of central importance in the century to follow: his concern with unifying physical "worldviews" derived from fundamental natural processes and illuminated by scientific research; his fostering of theoretical physics as a vehicle for research into the workings of nature and hence, in the public's eye, as a provider of new, overarching worldviews; and the very character of the work and personality of the man, who at times seemed to an infatuated public much larger than life.

YOUNG EINSTEIN

Albert Einstein was no stranger to his times, but he made his own way in physics. Einstein was born of nonreligious Jewish parents in the southern German town of Ulm on March 14, 1879. Reportedly, as a baby, he did not talk until very late, perhaps not until after he had turned three. This has led to speculation that he had a disability or perhaps a mild form of autism, Asperger's syndrome. However, there is no solid evidence for this. His younger sister, Maria (or Maja), born 2½ years later, noted his desire to speak only in complete sentences, which may have encouraged his reticence to talk.[1]

Within a year of his birth Albert moved with his family to Munich. Albert's father and an uncle, an electrical engineer, opened a manufacturing firm for electrical apparatus in the Bavarian capital. The firm did quite well in the expanding market for new electrical devices, some manufactured under the uncle's own patent. Like other youngsters of the period, Albert enjoyed a fascination with technology, and he readily participated in the family enterprise. The future theorist apparently owed to this period his lifelong interest in technical gadgets and inventions, several of which he later patented and published himself.[2]

Albert entered a public elementary school in Munich, where the majority religion, in this case Catholicism, was customarily taught. Although the parents were not religious, they hired a tutor to provide private lessons in Judaism. At the same time Albert began taking violin lessons and later became an excellent violinist, an activity he enjoyed the rest of his life. Religion, music, and nature began to intermingle in his thinking.[3] As Einstein entered adolescence, he began to wonder about the ultimate meaning of life. For answers, he turned first to religion: "Thus I came—despite the fact that I was the son of entirely irreligious (Jewish) parents—to a deep religiosity."[4] The religiosity, however, did not last long, for his reading of popular science books at the time convinced him that "the stories of the Bible could not be true." "The consequence," he wrote in 1949, "was a positively fanatic (orgy of) freethinking coupled with the impression that youth is intentionally being deceived by the state through lies; it was a crushing impression. Suspicion against every kind of authority grew out of this experience, a skeptical attitude towards the convictions which were alive in any specific social environment—an attitude which has never again left me, even though later on, because of a better insight into the causal connections, it lost some of its original poignancy."[5] Einstein thus refused to be bar mitzvahed, and he always maintained skepticism toward all forms of authority—scientific, social, and otherwise. Yet adherence to belief in a supreme intelligence continued throughout his life, but it was a belief similar to that of the eighteenth-century rational philosopher Baruch Spinoza: belief in an impersonal, cosmic, scientific intelligence that guaranteed a rationality to nature that was susceptible to human inquiry. "Like Spinoza," wrote philosopher Isaiah Berlin, "[Einstein] conceived God as reason embodied in nature, as being, in a literal sense, a divine harmony."[6]

The Einstein family business suddenly failed after the Einsteins lost a municipal electrification contract in Munich to the German electrical firm Siemens. In 1894 the family pulled up stakes and moved to Milan, in northern Italy, where business prospects proved brighter, but they left Albert, then aged fifteen and in his sixth year of *Gymnasium* (high school), behind to complete his schooling.

Boarding with an aunt, the teenager lasted alone in Munich only half a year. He quit school when Christmas vacation arrived in December 1894 and headed south to join his family. The antiauthoritarian teenager later complained of the forced teaching methods and military demeanor of his teachers, but the prospect of military obligations if he remained in Germany past his sixteenth birthday surely also encouraged retreat. He quickly renounced his German citizenship. Later he acquired Swiss citizenship, but to his likely relief he was exempted from Swiss army duty owing, the records indicate, to varicose veins and sweaty, flat feet.[7]

Upon arriving in Milan, the young man assured his parents that he intended to continue his education. Though underage and without a secondary-school diploma, Albert prepared on his own to take the entrance examination to the Swiss Federal Polytechnic, better known by its German acronym, ETH. Deficiencies in the nonscientific subjects apparently doomed his exam performance, but this resulted in his fortunate placement for a year in a nearby cantonal school to complete his secondary-school studies. Boarding in the stimulating home of one of his teachers, the new pupil blossomed in every respect within the liberal environment of Swiss education and democracy. The sojourn even inspired a brief but passionate romance between Einstein and one of the teachers' daughters, Marie Winteler, which ended soon after graduation.[8] (Maja, who followed Einstein to the school and later received a degree in romance languages, married Marie's brother.)

Einstein earned high marks at the cantonal school, graduated in 1896, and entered the teacher training program at the ETH, headed for certification as a secondary-school math and physics teacher. The four-year course of study entailed a rather sophisticated level of mathematics and original laboratory research in the field of the ETH's main physics professor, H. F. Weber, noted for his contributions to electrical engineering. But training in the contemporary fields in which Einstein would later make his mark was conspicuously absent; he mastered those subjects entirely on his own.[9]

Matters grew difficult after graduation. First, there was a problem regarding Mileva Marić, a young Serbian woman who had come to

Zurich to study physics at the ETH and was in the same year as Einstein. As indicated by recently discovered letters that have been widely publicized and explored in the popular literature, a romance arose between Mileva, the only woman in Albert's class, and the handsome and witty future science teacher. The letters written to each other when they were apart reveal their romance and, more importantly, their studies together. "I'm so lucky to have found you," he once exclaimed, "a creature who is my equal, and who is as strong and independent as I am!" Sometime in the year of Einstein's graduation Mileva inquired, "Because I like you so much, and because you're so far away that I can't give you a little kiss, I'm writing this letter to ask if you like me as much as I do you?" The answer may have been, "I can hardly believe that we will be separated so much longer—only now do I see how madly in love with you I am!"[10]

Despite the Einstein family's vehement opposition to Mileva—she was not Jewish, she was crippled by a congenital hip defect, and she intended to become a professional scientist—the romance flourished. But when Marić discovered that she was pregnant in 1901, Einstein began to distance himself. His girlfriend gave birth to a daughter early the following year, whom they named Liserl ("little Lise"), but Einstein never saw her, and no trace has been found of her, despite strenuous search efforts. Returning to Serbia, Marić may have given the child up for adoption, or she may have died early of illness. Not until later did Einstein's family, which continued to oppose any talk of marriage, finally accede to the liaison.

Einstein's concerns were apparently elsewhere at the time—on his work and on the problem of his career, which was headed nowhere. For reasons that are unclear—probably anti-Semitism and Einstein's disrespect for professorial authorities—Weber worked against the young man's future. Albert was continually passed over for graduate assistantships. Defending her boyfriend, Mileva, too, ran afoul of Weber. Upset over the pregnancy, she flunked her final exams for the second time and left the ETH without a degree. For several years Einstein lived a discouraging existence of temporary teaching positions and freelance tutoring. Lacking an academic sponsor, Einstein's doctoral dissertation submitted to the University of Zurich did not succeed, and he withdrew the work in 1902. (He finally received his doctorate from Zurich in 1906.)

Two years after Einstein graduated from the ETH, his father died in Italy; but by then the young man's situation was on the upswing. Prompted by friends of the family, in 1902 the Federal Patent Office in the Swiss capital of Bern offered Einstein a full-time job as an entry-level patent examiner with the lowly title of "Technical Expert, Third Class." In addition, shortly before his death, Einstein's father had also given permission for his son to marry. Six months later, Albert and Mileva finally married, and in 1904 their first son, Hans Albert, was born. (Their second son, Eduard, was born in 1910; Mileva and Albert later divorced.) With a full-time civil service job, a wife and family, and a comfortable apartment in Bern, Albert Einstein's life would have seemed complete to most. For him, it had hardly begun.

The year following his first son's birth was for Einstein what some say 1666 had been for Isaac Newton, an *annus mirabilis*, or miraculous year. In March 1905 Einstein presented the first solid argument for the quantum structure of light: light, previously regarded as a wave, appeared to behave like a stream of invisible tiny particles or bundles, "quanta," of energy. In April, Einstein offered an explanation for the erratic motion of small bodies suspended in liquids (Brownian motion) that for the first time offered undeniable evidence for the existence of atoms; in June he submitted a paper titled "On the Electrodynamics of Moving Bodies," the special theory of relativity involving new conceptions of space, time, and mass; and in September he presented a formula that became perhaps the most famous equation in modern physics: $E = mc^2$, where E is the energy equivalent to an amount of mass m multiplied by the square of the speed of light, c.

How did he do it? More important for our purposes, what did each of these works mean, and why were they so significant? To answer these questions, we have to look back briefly at physics and Einstein's orientation to it at the turn of the century.

EINSTEIN AND THE MECHANICAL PROGRAM

Nature and Nature's laws lay hid in night:
God said, "Let Newton be!" and all was light.

—Alexander Pope

This couplet, penned by the poet Pope as an epitaph for Newton's tomb, perfectly encapsulated the wonder and delight of an age so enamored of Newton that it could attribute to him the discovery of the very secrets of creation. It was an age still illuminated by Newtonian science that gave rise to Einstein and his physics and conferred upon him the Newtonian mantle of public adulation. But by Einstein's miracle year, Newton's illumination of nature and nature's laws had already dimmed among scientists, tarnished, ironically, for the most part by the problem of light.

Newton's world was a mechanical and material one: a type of "clockwork universe" in which all natural phenomena arise from the properties of moving matter interacting incessantly with other moving matter. Newton believed that the moving matter consisted of small indivisible bits, or atoms, of matter existing in the vacuum of empty space. Their motion was measured against the background of a stationary, infinite, flat space, and in terms of a linear, steadily advancing march of time. The relationships among the speeds, accelerations, masses, and forces of all matter (atoms or aggregates of atoms) obeyed Newton's three laws of motion: those involving action and reaction, force and acceleration, and the propensity of matter to continue at rest or in uniform motion (unless acted upon by an unbalanced force) known as inertia.

To this simple, rational world of matter in motion Newton added his most famous and far-reaching, though essentially nonmechanical, achievement: the law of universal gravitation. For some inexplicable reason all matter by its very existence attracts all other matter in the universe, and it does so with a force given by a precise, mathematical relationship involving the masses and their distances from each other. This law—which applied equally to falling apples, to the orbiting moon and planets, and (we now know) even to galaxies scattered across the universe—remained for Newton an enormously successful yet frustratingly impenetrable fundament of nature. Given the empty vacuum between atoms and aggregates of atoms, how precisely does this force act across empty space to exert the pull of attraction on other matter? Indeed, how does one bit of matter "know" of the existence of other bits of matter on which to exert its force? Gravitation seemed to be a mysterious, nonphysical "action at a distance." Being part theologian, Newton turned in his private thoughts, as often occurs in such situations, to theological explanations.[11]

Nevertheless, the Newtonian, or "mechanical," worldview provided an understanding of nature and a research program that dominated physics until the early twentieth century.[12] This program called for the unification of all physics on the basis of Newtonian mechanics. In essence it entailed the reduction of all observed phenomena, even life itself, to mechanical foundations: motion, forces, and matter, preferably atomic matter, acting according to Newton's laws in absolute space and time. This ambitious program gained an enormous following by its many successes and applications, especially in such fields as chemistry and planetary science. But the successes were less astonishing and increasingly more contrived in the study of electricity and magnetism, which also seemed to operate by forces acting at a distance over empty space.[13]

During the course of the nineteenth century, Michael Faraday, James Clerk Maxwell, and Heinrich Hertz had formulated a quasi-mechanical theory of electromagnetic phenomena. According to this theory, electric and magnetic forces could not be treated, like gravitation, as some inexplicable "action at a distance" but as the effect of electric and magnetic fields existing in the space between electric charges. Since something had to transmit these energy-bearing fields between the charges, the fields were pictured as mechanical "lines of force" or strains in a hypothetical "ether" that filled all of space, a fluid or medium more subtle than ordinary matter. The ether was real enough, however, to exert forces on ordinary matter and to provide an impetus for certain mathematical relations among fields and charges known as Maxwell's equations.[14]

A leading problem for physicists by the last decade of the nineteenth century was to find a complete mechanical account of Maxwell's equations. The problem was made all the more acute by Hertz's confirmation that moving electromagnetic fields could actually break away from ordinary matter and propagate through the ether as an independent electromagnetic wave carrying energy, much like waves spreading on a pond from the place where a stone is thrown into the water. Visible light, Hertz showed, was one form of electromagnetic wave. "The incorporation of wave-optics into the mechanical picture of the world," Einstein later wrote, was a leading difficulty, and one that "was bound to arouse serious misgivings" about the mechanical program.[15]

What did Einstein know of these latest developments at the forefront of mechanical and electromagnetic theories? Plenty, it seems, and, through independent study, quite early. During his educational interlude in 1895, the sixteen-year-old demonstrated a remarkable command of traditional mechanical attempts to account for electromagnetic fields in an insightful essay, probably written for a relative, entitled "On the Investigation of the State of the Ether in a Magnetic Field."[16] From his recollections and his student letters to Mileva Marić, it is clear that Einstein further educated himself about, and was educated into, the overall mechanical worldview of that period. Throughout his studies at the ETH and afterward, Einstein read widely and intensively in the mechanical works of leading contemporary physicists and became fully committed at the time to the mechanical program. Indeed, his earliest scientific papers, published within a few years of graduation, concerned extensions of the atomic, mechanical point of view to several phenomena in physical chemistry.[17]

What is remarkable in Einstein's research is the motivation that unifying worldviews provided. In an essay published in 1940, Einstein wrote autobiographically of his "search for a foundation of the whole of physics": "From the very beginning there has always been present the attempt to find a unifying theoretical basis for all these single sciences . . . from which all the concepts and relationships of the single disciplines might be derived by logical process. . . . The confident belief that this ultimate goal may be reached is the chief source of the passionate devotion which has always animated the researcher."[18] Einstein, the researcher, started with an attempt to find a Newtonian mechanical foundation for all of physics, but by 1905 he realized that a new foundation would be required. As McCormmach and others have pointed out, the revolutionary works of Einstein's *annus mirabilis*, though seemingly so diverse in content and import, were actually united by Einstein's overall vision of physical reality, a vision that required the reform of the Newtonian, mechanical worldview in order to perfect, extend, and adapt it to challenges newly raised in that period.[19] Far from attempting a radical departure from "classical" Newtonian and Maxwellian physics, Einstein was attempting to resolve difficulties raised within them in nineteenth-century terms.

EINSTEIN AND THEORETICAL PHYSICS

Einstein's concern with a unifying worldview and his adherence to a research program derived from it were not unique to him or to physics. The "unifying spirit," as it was called, pervaded much of central European thought at the turn of the century. German idealism, neo-romanticism, and historicism, stretching from Immanuel Kant and G. F. Hegel (or even earlier) to Benedetto Croce and Wilhelm Dilthey, each pointed to some sort of transcendent higher unity, the existence of permanent ideas or forces that supersede or underlie the transient, ephemeral world of natural phenomena, practical applications, and the daily struggle of human existence. The scholar, the artist, the poet, and the theoretical physicist all strove to grasp that higher reality, a reality that because of its permanence and transcendence must reveal ultimate "truth" and, hence, serve as a unifying basis for comprehending, for reacting to, the broader world of existence in its many manifestations.[20]

Yet, like most historical periods, the turn of the century could sustain more than one major movement, and an alternative pointed in the opposite direction: toward the denial of mechanical explanations and transcendent laws, toward an emphasis on "positive" knowledge gained solely through analysis of observed data gained from painstaking empirical research. In an industrial age associated in the public eye with "mechanistic materialism," a rising industrial working class, and the threat of materialistic "scientific socialism" espoused by Marx and Engels, it is perhaps not surprising that many academics, most of whom belonged to the upper economic classes, hastened to embrace nonmaterial alternatives to so-called mechanism.

Historical and biblical criticism, the new philology, and every experimental science, including experimental physics, were among the beneficiaries of the "positivistic" movement. Among its most influential proponents in physics was physicist-philosopher Ernst Mach.[21] Starting with an empirically based critique of such fundamental concepts in mechanics as absolute space and time, Mach and the supporters of "empirio-criticism" rejected the very existence of material atoms and any attempts to understand them in terms of universal mechanical laws. But they did allow use of the fruitful concept of atoms if it enabled the most "economical" account of the phenomena.

Unlike Pope's Newton, Mach insisted that human understanding can never penetrate to the actual underlying mechanisms of observed phenomena, because our perception itself of nature is flawed by the limitations of our senses, through which all knowledge of nature comes. Atoms are not real entities but merely convenient constructs. Though influenced by Mach's critique of mechanics, especially regarding absolute space and time, Einstein later provided valuable evidence that atoms do in fact exist.

While European scientists and academics argued over research methods and the insights (or lack thereof) they provided, many European intellectuals, especially those in Germany, began to regard their research as a self-assigned service to the European cultural tradition. Because of the value society placed at that time upon cultural expression, their work became for them the definition and defense of a special upper-level social status to which they, whatever their individual social origins, felt entitled. Out of this social and intellectual mix arose the notion of a new educational elite and the so-called apolitical professor. Academic scholars argued that, in order to maintain their scholarly and scientific objectivity, they had to maintain their distance from ephemeral practical applications and the "dirty world" of political intrigue in which, by no coincidence, the working class was now increasingly engaged.[22]

As Europe succumbed to the outbreak of world war, Einstein became one of the first European professors to renounce the apolitical ideology and enter into the atmosphere of political struggle—with much criticism from many of his German colleagues. He also enjoyed laboratory work and the invention of practical gadgets, but in line with the thinking of his era and his own adherence to an intelligent cosmos, Einstein saw himself quite consciously as a physicist devoted not to empirical analyses and philosophical critiques but to the opposite endeavor—to the discovery of transcendent laws of nature and their consequences as revealed through theoretical research. He saw himself as a member of the new and small profession of theoretical physics.[23]

The work of the theoretician, then and now, entails the formulation of physical theories. This meant the study of experimental data, the invention of general hypotheses to account for the data through what Einstein later called "free creations of the human mind," selection of the most promising hypotheses through careful examination of the data, expression of the results in terms of quantitative or mathematical formulas, and the formulation of specific predictions and general conclusions.[24] Although theorists might engage in experimental research, their work as theorists does not include the actual gathering of data or the construction of experiments to test nature—what Galileo once called the "interrogation of nature"—nor does it concern practical or technical applications. But, at that time especially, it did include the uncovering of transcendent truths about the natural world that could be built into a general outlook about nature, a worldview of wide application both within and far beyond physical science. It was through this latter activity that Einstein and others exerted their most immediate influence on the broader culture.

Despite the experimental and technical orientation of his education, despite the secondary-school teaching degree it afforded, Einstein had evidently set his own long-range career course right from the start. In his final exam in French language at the Swiss cantonal school, Einstein submitted a revealing essay on "My Future Plans." In faulty French he indicated his intention to attend the Zurich ETH and to study math and physics for four years. After later obtaining a doctorate, "I imagine myself becoming a professor in these branches of natural science, choosing the theoretical part of these sciences. . . . It is only natural that one always likes to do those things for which one has a talent."[25]

What did it mean—intellectually and socially—for a young man to choose theoretical physics for a career at the turn of the century? Historians have devoted some attention to this question, but more needs to be known. From a purely statistical point of view, Einstein was entering a profession as rarefied in its numbers as it was in its subject matter. One international survey indicates that, fortunately for German-speaking Einstein, theoretical physics in 1900 was primarily a German academic affair. With sixteen full-time faculty members in theoretical and mathematical physics, Germany possessed twice the number of faculty members in those fields as her closest competitors, Italy and Austria-Hungary. In the same year, Switzerland and the United States possessed only three faculty members apiece in the same disciplines. Ten years later, Germany still led with sixteen, Switzerland held steady with three, but Italy had jumped to twelve, while the United States fell even further behind, unable to catch up until the 1930s.[26]

Yet even in the homeland of German idealism, theoretical physics was still considered a second-rate profession. At the turn of the century, physics everywhere still meant primarily experimental physics. "Physics is an experimental science!" one Harvard professor exclaimed.[27] Mathematical physics—the study of equations of use to physics research—was considered an ungainly hybrid. Because it required no experimental equipment, theoretical research often served as a temporary holding position for junior faculty until they could move up to a full professorship with access to a laboratory.[28] Germany's enormous lead in theoretical-physics faculty actually represented only 15 percent of the total German physics profession. But this relatively high fraction reflected not only the career difficulties encountered by junior faculty but also the efforts of abstract mathematicians at the time to bring their work closer to practical problems. That more German theorists were Jewish than were members of other academic professions has been attributed to the greater opportunity afforded Jews in this then less desirable occupation.[29]

Anti-Semitism apparently contributed to Einstein's dire job situation after graduation: the independent researcher complained in his early papers that his lack of access to a laboratory prevented his confirming the results he had derived theoretically. Yet access to a laboratory would probably not have rendered Einstein an experimentalist; his commitment to theoretical physics lay much deeper than mere job circumstances. His understanding of the aims of his profession provided a crucial difference between himself and many of his experimentalist contemporaries.

Like his good friend and colleague, distinguished Berlin theorist Max Planck, Einstein viewed theoretical work as a high calling, not unlike that to the priesthood, a calling to search for truths and absolutes that survive the personal and transitory realities of everyday life. This was a calling that differed from experimental and even from mathematical physics. Pure mathematics was no temptation, Einstein later wrote, because he felt he possessed no intuitive sense for the important problems. The fundamentals were the objects of his interest, and in theoretical physics, he wrote, "I soon learned to scent out that which was able to lead to fundamentals and to turn aside from everything else, from the multitude of things which clutter up the mind and divert it from the essential," and, "Out yonder there was this huge world, which exists independently of us human beings and which stands before us like a great, eternal riddle. . . . The contemplation of this world beckoned like a liberation, and I soon noticed that many a man whom I had learned to esteem and to admire had found inner freedom and security in devoted occupation with it."[30]

One of those devoted men was Max Planck, who shared Einstein's goal not just of comprehending the world "out yonder" but of discovering nature's laws and even of achieving the ultimate goal:

a simple, elegant, all-encompassing view of nature that would unite all of physical science. "It has always seemed to me," wrote Planck in 1918, "that the most important thing, the goal that guided all of my scientific endeavor, is the greatest possible simplification and unification of the physical worldview and that the first means of reaching this goal is the reconciliation of opposites through mutual fertilization and amalgamation."[31] Confronted with challenging dualities and difficulties within the reigning worldviews at the turn of the century, Einstein sought such a unification in 1905, and throughout his life. Max Planck, co-editor of the German journal in which Einstein published all of his important early works, would act as Einstein's earliest and strongest supporter.

TECHNICAL EXPERT, THIRD CLASS

I f Einstein held to the nineteenth-century ideal of unifying physics on the foundation of mechanics, by the turn of the century he was in a dwindling minority. Most other theoretical physicists sought unity in one of two nonmechanical alternatives: the so-called energetic or electromagnetic points of view. These alternatives arose from nineteenth-century challenges to the mechanical program in studies of heat and electromagnetism. It was in an effort to reform mechanics and electrodynamics in the wake of these developments that Einstein produced his 1905 works.

THE CHALLENGE OF HEAT

The study of the dynamics of heat flow, or thermodynamics, had culminated in two fundamental laws regarding heat. The first law related heat, energy, and useful work to each other in thermal processes. This law could be understood in terms of the motions and collisions of Newtonian atoms. The second law, however, could not. According to the second law, the flowing of heat in natural processes, such as the melting of an ice cube, is always irreversible; that is, heat will not naturally flow of its own accord in the opposite direction—the melted cube at room temperature will not refreeze by itself. How to account for this in mechanical terms?

If, as Newton and others had suggested, all matter consists of atoms (or molecules), then heat is nothing but the energy of motion, or kinetic energy, of the atoms. But, like so many bouncing marbles or billiard balls, all atoms in their microscopic interactions must obey Newtonian mechanics. Those interactions *are* reversible: a motion picture of a collision between simple atoms will look perfectly normal if it is run backward in time. So how does the irreversibility of macroscopic events, such as melting ice cubes, arise?

This and other paradoxes encouraged those who, like Ernst Mach, chose to deny the very existence of material atoms. One group, led by physical chemist Wilhelm Ostwald, seeing their chance in paradox, rejected the entire mechanical program, holding the laws of thermodynamics, not mechanics, as fundamental.[1] Mechanics required hypotheses about matter and invisible atoms in motion, but thermodynamics referred only to energy and its observed transformations in the everyday world. Because thermodynamic laws were closer to laboratory observations, universal, freed of paradox, and independent of matter, Ostwald and his followers proclaimed the predominance of a new "energetic" worldview: energy and the laws of thermodynamics are the bases for understanding all processes within physical science, and even beyond. Upholders of this view, known as "energeticists," though unable to make much of their position, maintained it even into the depths of World War I, which they condemned as a wanton waste of energy (to say little of human lives).

Others, of course, held tightly to material atoms. They found support in the work of Maxwell, Rudolf Clausius, and Ludwig Boltzmann, who managed to resolve the reversibility paradox in favor of atoms. The second law of thermodynamics says that most natural processes are irreversible, in contradiction to the Newtonian mechanics of atoms. Boltzmann in particular resolved this contradiction by interpreting the second law as a new type of law: a statistical law, not an absolute one. Since there are so many atoms or molecules, even in a tiny ice cube, it is extremely unlikely—but not impossible—for the myriads of molecules in a melted cube to return in a finite time from the disorder of a liquid to their original orderly, crystalline arrangement. The macroscopic properties of heat and material objects, such as irreversibility, arise from the statistical behavior of numerous mechanical atoms, a behavior to be described by a new "statistical mechanics."

Boltzmann and US physicist J. Willard Gibbs provided the first accounts of how exactly the second law of thermodynamics arises from the statistical behavior of myriads of randomly moving atoms. Unaware of these writings, Einstein devoted three brilliant early papers during the years 1902–1904 to an independent derivation of the second law in the course of developing his own statistical mechanics, based on atoms and mechanics. Continuing in this work, Einstein used mechanics, atoms, and statistical arguments to achieve what he called a "general molecular theory of heat," confirming that both laws of thermodynamics are, indeed, fully explicable on mechanical grounds.[2]

BROWNIAN MOTION

In his doctoral dissertation, submitted to the University of Zurich in 1905, Einstein developed a statistical molecular theory of liquids; then, in a separate paper, he applied the molecular theory of heat to liquids in obtaining an explanation of what had been, unknown to Einstein, a decades-old puzzle. Observing microscopic bits of plant pollen suspended in still water, English botanist Robert Brown had noticed in 1828 that the pollen seeds exhibited an incessant, irregular "swarming" motion, since called "Brownian motion." Although atoms and molecules were still open to objection in 1905, Einstein predicted that the random motions of molecules in a liquid impacting on larger suspended particles—such as pollen seeds— would result in irregular, random motions of the particles, which could be directly observed under a microscope. The predicted motion corresponded precisely with the puzzling Brownian motion! From this motion Einstein accurately determined the dimensions of the hypothetical molecules.[3]

By 1908 molecules could no longer be considered hypothetical. The evidence gleaned from Brownian motion on the basis of Einstein's work was so compelling that Mach, Ostwald, and their followers were thrown into retreat, and material atoms soon became a permanent fixture of our understanding of our physical world. Today, with the advent of scanning tunneling microscopes, scientists essentially observe and even manipulate individual atoms for the first time—a circumstance that would satisfy even the most entrenched Machian skeptic.

In the course of his fundamental work on applications of statistical methods to the random motions of Newtonian atoms, Einstein discovered a connection between his statistical theory of heat and the behavior of electromagnetic radiation—the first step toward his hoped-for unification of these two fields. Using his statistical theory of heat, Einstein obtained a mathematical expression for the fluctuations, or sporadic oscillations, in the average energy of any system. He applied this expression to the average energy of thermal radiation—the electromagnetic waves given off by glowing bodies—in a perfectly reflecting box (often called a "black body"). He obtained results in close agreement with experimental observations. This connection, he declared in obvious understatement, "ought not to be ascribed to chance."[4] For a physicist such as Einstein, interested in uniting perspectives, the connection provided an extraordinary opportunity. Einstein's fundamental papers on relativity and quantum theory, also submitted in 1905, may be seen as far-reaching explorations of the connection. They arose, however, in the context of the new electromagnetic theory of electrons.[5]

ELECTRON THEORY

As noted in the previous chapter, Faraday, Maxwell, and Hertz had produced a theory of electromagnetic phenomena based upon the transmission of electromagnetic effects between charges by a hypothetical medium, the ether. This theory led to the important discovery that the fields can break away from charges and propagate through the ether by themselves, forming a so-called electromagnetic wave, of which visible light is one form. (Invisible electromagnetic waves include radio waves, x-rays, microwaves, and many others.) The electromagnetic waves spread through the massless ether like rippling waves across the surface of a still pond.

While the exact mechanical workings of this "ghostly" ether remained a major puzzle to Einstein and many others, the problem of electric charge and the discovery of the electron in 1897 brought matters to a head. The electron posed a riddle. It possessed the smallest mass known, and it carried the smallest electric charge, the electron charge. Because of its charge, it was the source of electromagnetic fields and was in turn subject to electromagnetic forces. Yet as a finite particle, a speck of charged matter, it differed by its very nature from the continuous, massless ether of electromagnetic theory. It simply did not fit into existing electromagnetic theory.

During the 1890s and the early years of the twentieth century, Dutch physicist H. A. Lorentz succeeded in formulating a new "electron theory" that amalgamated the electron with Maxwell's electromagnetic ether. Lorentz could attribute most of the properties of an electrically charged piece of matter to electromagnetic forces propagated within the ether and exerted by the ether on electrons within the piece of matter.

The enormous success of Lorentz's electron theory, together with the failure of every attempt to reduce Maxwell's equations to mechanical foundations, spawned a new, nonmaterial alternative to the mechanical worldview, the electromagnetic view of nature, which captured the imaginations of many physicists. According to this worldview, electromagnetic fields, not matter, are fundamental; Newtonian mechanics—along with all physical phenomena—

would be reduced to electromagnetic interactions carried by the ether between electrons. Even mass itself, starting with the mass of the electron, would somehow arise from electromagnetic fields. This last notion, a radical goal even for those times, received important support when Walter Kaufmann made the startling discovery that the mass of the electron is not fixed and unchanging, as everyone thought, but actually increases as the electron picks up speed in an electromagnetic field.

Despite the successes of the electromagnetic worldview, numerous problems hindered its universal triumph. Foremost among them was the relationship of the electromagnetic ether, the supposed carrier of forces and fields, to matter that had mass but no electric charge. When matter moves relative to the ether, is the ether pulled along with it, or does it remain stationary? Also, what are the origins of irreversibility when a hot body emits electromagnetic radiation? In addressing such questions, Einstein extracted from Lorentz's electron theory the research area in which he worked and the problems that he would solve in formulating his famous contributions of 1905.[6]

MILEVA'S CONTRIBUTION?

The relative motion of matter and ether had posed a problem for years, and, Einstein wrote his friend Marcel Grossmann as early as 1901, one with which he was particularly occupying himself.[7] One historical question that has arisen recently concerns the extent to which Einstein's famous works of 1905 derived from discussions with his friends and, especially, with his fellow physics student and future wife, Mileva Marić. The many letters recently published from Einstein to Marić, along with a few responses from Marić to Einstein, dating from 1897, offer intriguing insights into this question and into the early genesis of Einstein's papers.[8]

In his letters to Marić, Einstein occasionally outlined in very brief terms his latest research, including his early work on the relativity and quantum theories. The letters to her indicate further that

Einstein was also discussing his work with his other friends and with his doctoral advisor, Professor Alfred Kleiner at the University of Zurich. But in one letter to Marić in 1901 Einstein looks forward to the time "when we are together and can bring our work on relative motion to a successful conclusion." Later that year Einstein seemed to be looking forward to a continuation of their joint work after their marriage: "When you're my dear little wife we'll diligently work on science together so we don't become old philistines, right?"[9] But there is little else regarding the nature of their work together. In other instances Einstein breaks off the discussion of his ideas in the expectation that they will discuss the matter further when they meet. Marić's few surviving responses contain no science.

These tantalizing yet frustrating letters have spawned a debate over Marić's role in Einstein's work. Obviously he discussed his work with her, and she was apparently able to follow it. The editors of Einstein's papers have suggested that she served as a sounding board for his ideas and that the early draft paper on relative motion probably differed greatly from the final relativity paper four years later.[10] Other writers have argued that Marić surely made a contribution that went unacknowledged by Einstein. After all, he referred in the above quotations to "our work on relative motion" and seemed to be promising continued collaboration after their marriage. He had also fruitfully discussed his ideas at various times with a number of other people. Yet in the published relativity paper of 1905 Einstein thanked only "my friend and colleague M[ichele] Besso . . . for several valuable suggestions."[11] Was Einstein, deliberately or not, suppressing the contributions of others to his work? Did their contributions justify an acknowledgment by Einstein? In the case of Marić, the person closest to Einstein and the one probably most privy to his work as it developed, was this another sad example of a successful man suppressing the recognition due the woman who had helped to make him successful?

John Stachel, Einstein scholar and former Einstein editor, argues strongly against the views that Marić made any significant contribution.[12] Marić herself never claimed during her lifetime that any of the ideas in his papers were hers. She always spoke of "his work" and "his research," and she never published any research on her own, whereas Marie Curie, for example, continued to publish after her husband's death. But Marić was no Curie. After she failed her exams, married Einstein, and began bearing his children, she probably abandoned any hope of her own career.

Andrea Gabor and others have suggested that, given Einstein's self-proclaimed weakness in mathematics and Marić's ETH education, she may well have helped him with the mathematics of his papers and may have added new insights as she did so.[13] But, as Stachel points out, mathematics was for Marić a stumbling block on the ETH exams. At the same time, Einstein's proclaimed weakness in mathematics actually concerned only the less common and more abstract areas, such as those needed later for general relativity. The mathematics of the 1905 papers was familiar to most ETH students and well within Einstein's capabilities.

We may never know for certain the answers to these questions. However, in view of the intensity of Einstein's concern with the problems of 1905, extending back to well before he met Marić, and the originality and idiosyncratic nature of Einstein's relativity and quantum papers, it seems unlikely that others could have made fundamental new contributions to these works. At best, the others probably acted as sounding boards for his ideas, and it was probably Besso who performed the greatest service in that regard. Yet it also seems likely that Marić did discuss his ideas with him and read through the manuscripts and galley proofs of his papers before publication, looking for errors and at times perhaps even offering editorial comments for which some acknowledgment may have been due.

Einstein and Marić were married on January 6, 1903, and they remained close until the success of his 1905 papers began to pull him in other directions. There was no need for correspondence after they moved into their Bern apartment. Only if further evidence arises can we know for certain what, if any, contribution Marić made to Einstein's papers during the "miraculous year" of 1905.

SPECIAL RELATIVITY

Maxwell's electromagnetic equations were only valid for matter that did not move relative to the ether. But what happens when matter moves through the ether like a fish through water or a car through still air? Are the equations changed? And does the ether remain at rest, or is it "dragged" along by the matter? Early experiments on light, considered as waves in the ether, seemed to indicate a partial dragging, whereas Lorentz had built his theory of electrons on the assumption that the ether remained absolutely at rest as matter moved through it.

Neither view could be supported by the famous Michelson-Morley experiments, conducted in Cleveland in the early 1880s. These experiments marked an early appearance on the world stage of a new community of researchers that for a generation would be especially known for its experimental studies of light. By measuring the speed of light in different directions as the earth moved rapidly through space on its annual orbit around the sun, Albert Michelson and Edward Morley expected to observe differences in the speed of light in different directions. Like still air rushing over a car as it speeds down a highway, an "ether wind" should pass over the earth as it moved through the ether, whether stationary or partially dragged. Since the ether carried the light waves in Michelson's and Morley's apparatus, this "wind" should yield different speeds of light in different directions. The results were unsettling: Michelson and Morley could detect no ether wind at all—a contradiction of every prevailing theory![14]

The resourceful Lorentz and G. F. FitzGerald, an Irish mathematician, independently came up with an ingenious answer, the famous Lorentz-FitzGerald contraction. They suggested that when objects, such as the apparatus used by Michelson and Morley, move relative to the ether, the ether somehow contracts their molecules slightly in the direction of motion. This contraction is just enough to cancel out the detection of any ether wind and undercut any experiment such as Michelson's and Morley's.[15]

Lorentz incorporated the undetectable ether wind into his theory by formulating a way to mathematically transform the

Maxwell equations, which describe the propagation of light, from a stationary viewpoint (the ether) to one moving relative to the ether (the earth). This required transformations of the quantities representing distance and time in the equations, such that—up to the accuracy of the Michelson-Morley experiments—Maxwell's equations looked the same to every observer. The point of this exercise was that, if Maxwell's equations appear the same to every observer, then the measured speed of light will be the same, and no ether wind can be detected in Cleveland or anywhere else, whether at rest or in motion relative to the ether. On the whole, however, Lorentz could not justify the form of his equations or the supposed contraction. These were purely ad hoc inventions, taken as it were from nature to describe what is seen without any deeper justification or derivation—hardly a satisfactory solution.[16]

Although Lorentz's responses to the problem of relative motion with respect to the ether were quite new, the problem of relative motion was as old as humankind. Is the rising of the sun in the east every morning due to the motion of the sun away from the stationary horizon of the earth, or to the fall of the horizon away from the stationary sun as the earth rotates on its axis? Solely on the evidence of a rising sun, both perspectives can be regarded as equally valid; there is no way to decide between them. In order to point this out and thus to enable serious consideration of the idea of a rotating earth and a stationary sun, Galileo in the seventeenth century formulated the first "relativity principle," removing the distinction between moving and stationary observers. A cannonball falling from the top of a ship's mast, he noted, will land at the base of the mast, regardless of whether the ship is at rest or in motion with uniform velocity. People on the ship will not be able to distinguish their own state of rest from uniform motion by a falling ball or any other such mechanical experiment performed within the "reference frame" of their ship. The same held for a person on a rotating and orbiting earth, Galileo argued. Despite our prejudices for a stationary earth, we on the earth cannot tell (without sophisticated experiments) if we are really at rest or moving with the rotation of the earth each day and its annual orbit around the sun.

Galileo's relativity principle referred only to mechanical experiments, such as falling cannonballs on moving ships. The Michelson-Morley experiment and Lorentz's modification of his theory suggested to French mathematician Jules-Henri Poincaré an expanded relativity principle. If light and Maxwell's equations behave the same in reference frames at rest or in uniform motion with respect to the ether, then, again to the limit of accuracy of the Michelson-Morley experiments, Galileo's relativity principle should be extended to include optical phenomena as well. In the years from 1900 to 1904, Poincaré enunciated this experimentally based relativity principle several times and called upon Lorentz to incorporate the principle into electron theory as the rationale for introducing his ad hoc equations for transforming between viewpoints at rest and in motion relative to the ether (described above). Lorentz apparently resisted this approach as in itself a mere ad hoc assumption.[17]

Instead, in order to resolve these and other difficulties within a single coherent theory, one year later, in 1905, Einstein reformed electron theory in a paper entitled "On the Electrodynamics of Moving Bodies."[18] Einstein's paper, the foundation of the theory of relativity, contained some of the exact same language and results as those of his predecessors: the relativity principle; Lorentz's transformation equations; and even the notion that light propagates at a constant speed regardless of the motion of the source or the observer, what some physicists had seen as a consequence of light carried by a stationary ether filling all of space. But this was no mere repetition of the work of his predecessors. It was a new and revolutionary theory that simultaneously encompassed and transcended the past.[19]

Because of the identity between Einstein's results and those of Lorentz and Poincaré, Einstein's contemporaries experienced difficulties at first in distinguishing Einstein's new "relative theory"—or, as it became known, his "theory of special relativity"—from the electron theory of Lorentz and Poincaré. Some, later driven by animosities fed by irrational motives, refused to recognize any distinctions. Even as late as 1953, Sir Edmund Whittaker, a prominent mathematical physicist and historian, published the second volume of his *History of the Theories of Ether and Electricity*, in which he presented

Einstein's new theory in a chapter entitled "The Relativity Theory of Poincaré and Lorentz."[20] Whittaker's startling portrayal helped launch a period of intensive historical research that culminated in the 1970s in a much deeper comprehension and appreciation of Einstein's contribution. This is still the most thoroughly researched facet of Einstein's life and work.

On the basis of that flurry of research, the most significant difference that emerges between the work of Einstein and that of Lorentz and Poincaré lies in the distinction that Einstein himself later drew between "principle theories" and "constructive theories": those that are built, respectively, from the top down (e.g., that of Einstein) or from the bottom up (e.g., that of Lorentz and Poincaré).[21] Constructive theories are built upon hypotheses and equations extracted from experimental evidence. Using Poincaré's empirical relativity principle and Maxwell's and Lorentz's equations, Lorentz and Poincaré had sought to *construct* a new electron theory from the ground up to meet the challenges posed by relative motion and the missing ether wind.

Einstein worked in reverse. So-called principle, or deductive, theories start from general principles or postulates that are assumed universally valid, even in situations where no experimental evidence is available. Conclusions are then derived from the principles and compared with actual data. Contrary to the constructive approach of Lorentz and Poincaré, in 1905 Einstein posited two universally valid (not empirical) fundamental postulates: the relativity principle and the constancy of the speed of light. Then he *derived* the consequences. The consequences, though obtained through an entirely different approach, turned out to be quite familiar. Among other things, the Lorentz-FitzGerald contraction reappeared—but now with new origins, sounder justification, and different implications. For Einstein, the relativity principle was not just an empirical hypothesis but a universal principle; his reform of electrodynamics was not just intended for electron theory but for all of physics, including mechanics; and his motivations were not merely practical and empirical but fundamentally aesthetic and conceptual, aimed at achieving elegance, simplicity, and a unified worldview.

For Einstein, who previously held only to the mechanical view, the mechanical and the electromagnetic worldviews were now to exist on an equal footing, such that clues from mechanics could be used to reform electrodynamics, and vice versa. One of the clues arose from considerations of symmetry, or balance. If the basic features of mechanics were to remain valid, Galileo's relativity principle in mechanics should apply also to electrodynamics. That it did not yet apply to electrodynamics, Einstein emphasized in so many words in the opening paragraphs of his paper, was evident from the "asymmetry" in the way Maxwell's theory treated, for example, a conducting wire moving near a magnet differently from a magnet moving near a conducting wire.

Einstein's elevation of the relativity principle to a universal postulate brings us to a second major difference between Einstein and his predecessors: the ether. Although it is still uncertain what the Bern patent clerk knew of the Michelson-Morley experiments, which had so motivated others, Gerald Holton has pointed out that it would have made little difference, anyway.[22] Einstein had utilized the ether earlier and gave evidence of having struggled for several years with the ether-based electron theory. But in the end, armed with the relativity principle and the constancy of the speed of light, he simply dismissed the ether in a few sentences as "superfluous," and with it, he wrote, went the familiar "idea of absolute rest."[23] Almost everyone since Newton (Mach was a major exception) had regarded the ether, or at least space itself, to be at rest and all motion to occur against the backdrop of this state of absolute rest. Instead, without a reference point at rest, all motion becomes relative and every concept involving space and time must be reconsidered in relative terms—a mind-boggling situation for physicists and for many others. Whereas Lorentz and Poincaré struggled to preserve the ether as a reference frame at rest, Einstein dispensed with absolute rest and placed electromagnetic fields on their own—propagating as independent physical entities throughout an etherless empty space. As such, electromagnetic fields were not reducible after all to mechanical foundations.

Further differences between Einstein and Lorentz have been

listed by Stanley Goldberg.[24] To Lorentz, the Lorentz-FitzGerald contraction was an actual, physical deformity of the molecules within moving objects. To Einstein, it was an unavoidable artifact of a measuring process that involves two observers in motion relative to each other; it arose from the measurement process itself. Like two persons observing Leonardo da Vinci's *Mona Lisa* from different angles, their different perceptions of her enigmatic smile arise only from their different perspectives. If Lorentz's theory is a theory of electrons, Einstein's relativity theory is fundamentally a theory of measurement. But this theory has something profound to say about all of physics: because of the relativity principle and the constancy of the speed of light, the laws of physics must appear to be the same to all observers, regardless of their uniform motions relative to each other. That can only happen if the fundamental, measured quantities that enter into these laws and equations—time, length, and mass—are different to observers moving at different speeds relative to the object to be measured.

The consequences of Einstein's theory of special relativity—special because it refers only to uniform velocities—are strange and confusing, even for specialists. Similar reactions occurred centuries earlier when Galileo first enunciated his thought experiment involving the ship moving at uniform velocity. The cannonball falling from the mast of the moving ship would appear to a shipborne observer to drop straight down, but to an observer on the shore, it would appear to follow a parabolic trajectory, likewise landing at the base of the moving mast. So, which trajectory did the cannonball actually follow—the line straight down or the parabola? The answer is, both—and neither. Each observer's observation is valid in its own reference frame, yet each is no more than an artifact of the measurement, or observation, undertaken by that observer.

Einstein extended this strange situation to a study of exactly how space, time, and mass differ for different observers in motion relative to each other. Ever since Newton, these basic elements of most measurements had been taken as absolute "givens" in any theory. Like the observers of Galileo's cannonball, in relativity theory an observer at rest relative to a clock, a meterstick, and a ball of mass would measure each to possess the expected values. But if that observer and all of his apparatus are moving at uniform velocity relative to another observer at rest, this latter observer would see a contracted meterstick, a slowed clock, and a ball with increased mass (see fig. 1). As calculated using Einstein's equations, each of these effects would become more pronounced as the observer's speed increases until, at the speed of light, about 186,000 miles per second (300,000 km/sec), an extreme state is reached. Length shrinks to zero, time stops, and mass becomes infinite! For these reasons, the speed of light is often considered the "speed limit" of the universe.

But there is more. In reforming electron theory, Einstein was attempting to bring it into closer alignment with a relativistic Newtonian mechanics. Rather than absorbing mechanics into electrodynamics, as others would have it, Einstein developed a new "kinematics" of motion, mass, space, and time upon which to base a new electrodynamics of moving bodies. In unifying mechanics with electromagnetic theory, he believed, a new relativistic physics would be achieved that would transcend both. At the end of 1905 Einstein announced in a separate paper that he had found important evidence for the new unification. Exploring the consequences of his equations, he had discovered that electromagnetic radiation, like matter, can carry inertia, or mass. In fact, a given amount of electromagnetic energy is equivalent to a certain amount of mass. The equivalence is controlled by a very large number, the square of the speed of light: a little mass represents an enormous amount of energy. Expressed in more modern notation as the equivalence of a certain amount of energy to a given amount of mass, and vice versa, this relationship may be written $E = mc^2$.[25]

The connection between mass and energy recalls an earlier connection: the one between mechanics and thermal electromagnetic radiation that Einstein had discovered in 1904. Einstein explored these connections in a fourth paper presented in 1905, "On a Heuristic Point of View Concerning the Production and Transformation of Light," a paper that included Einstein's revolutionary light-quantum hypothesis.[26]

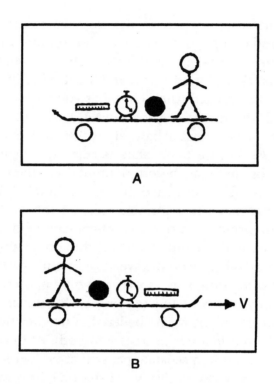

Figure 1. Relative motion. B is moving on a skateboard at constant velocity relative to A and to the ground, but B cannot tell this, unless he looks at A or the ground. It is a perfectly smooth ride. B could think that he is at rest and that A is moving backward at the same velocity that he (B) is moving forward. Thus A and B can each claim that he is at rest and that it is the other person who is moving.

QUANTUM THEORY

Light and other forms of electromagnetic radiation, such as radio waves, are obviously waves—or so everyone thought. Maxwell and Lorentz had firmly established the wave nature of electromagnetic radiation in electromagnetic theory. Numerous experiments on the interference, diffraction, and scattering of light had confirmed it. We can well appreciate the shock and disbelief when Einstein argued in 1905 that under certain circumstances light behaves instead not as continuous waves but as discontinuous, individual particles. These particles, or "light quanta," each carry a "quantum," or fixed quantity, of energy, much as automobiles produced by an assembly plant arrive off the assembly line as individual, identical cars—never as fractions of a car. The total energy of the light beam (or the total output of an assembly plant) is the sum total of the individual energies of these discrete "light quanta" (or automobiles), what are called today "photons." Theories of matter and electromagnetic radiation in which the total energy is treated as "quantized" are known as quantum theories. Although Einstein was not the first to break the energy of light into packets, it has been argued by Thomas S. Kuhn that he was the first to take this seriously and to realize the full implications of doing so.[27]

Like the special theory of relativity, Einstein's quantum hypothesis arose from an experimental puzzle and an asymmetry or duality in physical theories.[28] The duality consisted of the well-known distinction between material atoms and continuous ether, or, as Einstein wrote in the opening sentence of his light quantum paper, "between the theoretical concepts that physicists have formed about gases and other ponderable bodies, and Maxwell's theory of electromagnetic processes in so-called empty space."[29] As noted earlier, Boltzmann and others conceived of gases as consisting of myriads of individual atoms, whereas Maxwell and Lorentz envisioned electromagnetic processes as consisting of continuous waves. Einstein sought a unification of these two viewpoints by removing the asymmetry in favor of a discontinuous, "atomic," or quantum, theory of light. Resolution of an experimental puzzle encouraged this approach.

The puzzle concerned so-called black-body radiation, that is, the electromagnetic radiation given off by a hot, glowing coal in a fireplace, or the radiation emerging from a small hole in a perfectly black box containing electromagnetic radiation at a high temperature. Scientists at the German bureau of standards in Berlin, who were interested in setting standards for the emerging electric lighting industry in Germany, had measured the distribution of the total electromagnetic energy in a black box—which would also apply to a glowing light bulb—among the different wavelengths of the light. But no one until Max Planck, at the turn of the century, was able to give a single mathematical formula for the observed distribution of the energy among the emitted wavelengths. Starting with the Maxwell-Lorentz theory of radiation and some natural assumptions about energy, Planck hoped to derive this formula from the second law of thermodynamics. Planck failed to obtain the observed formula on these assumptions. Even Lorentz had to admit that his own electron theory could not account for black-body radiation.

Only by reluctantly introducing a radical new assumption into his mathematics could Planck attain the correct formula. The assumption was that the energy of the radiation does not act continuously, as one would expect for waves, but exerts itself in equal discontinuous quanta of energy. In essence Planck had discovered the quantum structure of electromagnetic radiation. But Planck himself did not see it that way; he saw the new assumption merely as a mathematical trick to obtain the right answer. Its significance remained for him a mystery. Kuhn claims that it is not to Planck in 1900 but to Einstein in 1905 that we owe the origins of quantum theory.

Encouraged by his brief but successful application of statistical mechanics to radiation in 1904, in 1905 Einstein attempted to resolve the duality of atoms and waves by demonstrating that part of Planck's formula can arise only from the hypothesis that electromagnetic radiation behaves *as if* it actually does consist of individual quanta of energy. The continuous waves of Maxwell's equations, which had been confirmed experimentally, could be considered only averages over myriads of tiny light quanta, essentially "atoms" of light.[30]

With his light quantum hypothesis Einstein could now account for certain hitherto inexplicable phenomena. Foremost among them was the so-called photoelectric effect: the ejection of electrons from a metal when irradiated by light. The wave theory of light could not yield a satisfactory account of this, since the energy of a wave is spread over its entire surface. Light quanta, on the other hand, acting like tiny billiard balls, could easily eject electrons, since the electron absorbs the entire quantum of energy on impact.

At first Einstein believed that the light-quantum hypothesis was merely "heuristic": light behaved only *as if* it consisted of discontinuous quanta. But in a brilliant series of subsequent papers in 1906 and 1907, Einstein used his statistical mechanics to demonstrate that when light interacts with matter, Planck's entire formula can arise only from the existence of light quanta—not from waves. The light quantum had to be real, even if its existence contradicted current theories of light. Einstein considered that these quanta, together with the equivalence of mass and energy, might result in a new electrodynamics based upon an atomic mechanics. But in 1907 he discovered that atoms in matter are also subject to a quantum effect.

Here he made use of another galling experimental problem. Experimentalists had found that as solid bodies are cooled, the amount of heat they lose failed to fit a simple formula that followed from Newtonian mechanics. Einstein showed that the experiments could be explained only on the assumption that the oscillating atoms of the solid lattice can have only certain, specific energies and nothing in between. In other words, even the motions of atoms—which are continuous in Newtonian mechanics—exhibit a quantum structure. Mechanics and electrodynamics both required revision, Einstein now concluded: neither could yet account for the existence of electrons or energy quanta.[31]

One culmination of Einstein's early work on both relativity and quantum theory occurred at the annual meeting of German (and German-speaking) Scientists and Physicians held in Salzburg in 1909. Einstein addressed the gathering as an "honored guest." The problem of light and the search for a unified physical worldview were as much on his mind as ever. His work on Brownian motion and statistical mechanics had served as extensions of the Newtonian mechanical worldview and a confirmation of the existence of atoms and molecules. The mechanical worldview had much in its favor, he believed, but it was not the last word. His reform of Lorentz's electron theory on the basis of the modified mechanics of special relativity theory would lead, he hoped by 1907, to a new "complete worldview that is in accord with the principle of relativity," what some called a new "relativity worldview," that would incorporate and transcend both mechanical and electromagnetic views.[32] But the quantum hypothesis, more than these previous works, pointed in a radically new direction: both mechanics and electrodynamics would have to be revised in order to account for the existence of the energy quantum in radiation and in matter.

However much Einstein sought to unify the foundations of physics, he was confronted again and again with persistent dualities. Returning in 1909 to the fluctuation formula that had served him so well in the past, Einstein applied it to electromagnetic radiation in empty space. He demonstrated that the fluctuation, or variation, in the energy of a mirror suspended in a black box full of electromagnetic radiation could be expressed mathematically as the addition of two independent terms. One term he attributed to the wave nature of the radiation; the other to its granular, quantum structure. As later became clear, he had brought into sharp focus one fundamental puzzle of twentieth-century physics: the wave-particle duality. Light can behave both as waves *and* as particles. The unification of mechanics and electromagnetism into one "physical world picture," as Planck called it, would entail a unification of the electron and the light quantum with the continuous character of the electromagnetic fields. The search for a "fusion" of the wave and particle aspects of electromagnetism—for a unified theory of disciplines of mechanics and electrodynamics—now became the goal and leading theme of Einstein's research to the end of his life.[33]

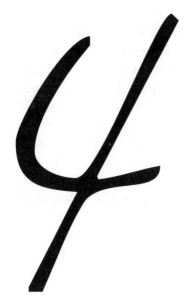

IMPACT ON PHYSICS

RELATIVITY

Einstein's publications on relativity and quantum theory exerted profound effects throughout physics on several levels: professional, intellectual, and methodological. In the process, Einstein's work transformed the nature and subject matter of physics, especially theoretical physics, into their influential, contemporary forms. But the effects of his work were felt only gradually, occurring in some cases over decades.

One early hindrance to the reception of Einstein's work among physicists was the enormous difficulty in grasping what Einstein had actually done. Both relativity and quantum theory met with initial incomprehension and even skepticism, but relativity theory received much more rapid and wider acceptance than did quantum theory. The impacts of these two theories on physics are therefore discussed separately.

THE RECEPTION OF RELATIVITY

Einstein's first paper on the theory of relativity appeared in the German research journal *Annalen der Physik* (Annals of Physics) in September 1905. Although misapprehension characterized its initial greeting everywhere, reception of the paper differed among nationalities. National traditions in science and the opinions of leading scientists in each nation proved the deciding factors during the first half-decade or so. One author succinctly characterized the responses of four nations in the chapter titles of his book: France: "As If It Never Happened"; Great Britain: "Defending the Ether"; the United States: "Defending the Practical"; and Germany: "When a Hundred Flowers Bloom."[1]

Responses among German-speaking scientists blossomed most extensively in the pages of German physics publications, but the reactions ranged at first from misunderstanding to rejection—not for technical but for conceptual reasons. Max Abraham and Fritz Hasenöhrl rejected the theory out of hand; Willy Wien and Max Planck experienced difficulty in accepting Einstein's postulate on the constancy of the speed of light (it seemed to contradict the relativity principle); and Hermann Minkowski, in formulating a mathematical representation of the theory, apparently failed to appreciate the demise of space and time as absolutes.[2]

Many physicists still believed that something had to carry the electromagnetic waves: if not the ether, then absolute space and time. Lorentz still preferred the ether. In a series of lectures presented at Columbia University in 1906, Lorentz declared in a discussion of Einstein's new electrodynamics: "Yet, I think something may also be claimed in favor of the form in which I have presented the [electrodynamic] theory. I cannot but regard the ether which can be the seat of an electromagnetic field with its energy and its vibrations, as endowed with a certain degree of substantiality, however different it may be from ordinary matter."[3]

Others saw in relativity theory a reactionary throwback to the old mechanical worldview, a worldview that many believed would be superseded at the forefront of research by the electromagnetic view. When Max Planck presented a lecture on an experimental test of the "relative theory" to the 1906 meeting of German Scientists and Physicians, a young scientist named Arnold Sommerfeld spoke for many when he declared, "In the question of principles formulated by Herr Planck I suspect that the men under forty years of age will prefer the electrodynamics postulate [the electromagnetic view], those over forty the mechanical-relativistic postulate. I give preference to the electrodynamic."[4]

Max Planck, an over-forty editor of the *Annalen der Physik*, did not yet prefer the mechanical-relativistic view, but he was the first established physicist to give Einstein's theory his full support. Einstein acknowledged in 1913 that this early support had proved crucial in gaining acceptance of the theory among his colleagues. Planck's first lecture on Einstein's paper at the University of Berlin in the fall of 1905 sent Planck's assistant, Max von Laue, to Bern on his next vacation to visit the theory's unknown author. Von Laue published the first textbook on relativity theory in 1911 and remained a lifelong friend and supporter. Planck himself worked on extensions of the theory and applications of it to experimental data.[5] As coeditor of the *Annalen*, he encouraged papers that offered new physical, rather than purely mathematical, insights into relativity theory. According to one count, during the years 1905–14 Planck presided over twelve dissertations at the University of Berlin on various aspects of relativity theory. Among them were Hans Witte's study of a mechanical foundation for electrical phenomena, Erich Hupka's measurements of the change in electron mass with velocity, and Erich Kretschmann's work on gravitation and relativity.[6]

Planck's early interest in relativity theory arose from its implications for the unified physical worldview that was the constant object of his research. Yet even as strong a supporter as Planck had difficulty comprehending the full implications of Einstein's work. In a 1906 paper on the principle of relativity, Planck mistakenly referred to this principle as "recently introduced by Lorentz and in a more general form by Einstein." He referred repeatedly to the "Lorentz-Einstein theory," in which "the principle has exact validity."[7] A year later Planck contemplated an experiment to measure the motion of the

earth relative to the ether; not until 1910 did he realize that the relativity principle rendered the ether and the notion of absolute rest
untenable. Yet even then he, like many other physicists, merely
replaced the ether with the concept of empty space. Even as late as
two decades after Einstein had published his original paper, some
physicists were still using the terms *ether* and *vacuum* interchangeably.

The enormous difficulty that Planck (and others) experienced in
comprehending what Einstein had actually done indicates the enormous transformation in thinking that the theory of relativity
required. Not all physical theories were constructed on the foundations of data and semiempirical hypotheses; this one was derived
from two universal principles, and its consequences were then compared with empirical data. Not all physical theories dealt with relationships among bits of matter moving through absolute space and
time; this one dealt with differing measurements of such fundamental constituents of physical theories as space, time, and mass
performed by different observers moving relative to each other. Not
all theories arose from sophisticated mathematical manipulations;
this one, though mathematical, rested upon conceptual reasoning
and fundamental principles, not equations. Because these principles
and the results derived from them were held to be universal, this
theory, also unlike most others, extended beyond the discipline
boundaries of electromagnetism and mechanics to place demands
upon all of physical science.

Nevertheless, no matter how difficult to grasp in all of its many
implications, the theory of relativity could not help but attract a
theoretical physicist such as Planck and, as a result, provide a new
definition and impetus to his emerging discipline. Despite his mis-
apprehensions, Planck saw in Einstein's theory what made theoretical physics what it was at that time: the search for a unified physical
worldview based upon invariants, universal principles, experimental
evidence, and conceptual intuition and reasoning, not just mathematical abstraction. The new theory greatly encouraged the spread
and success of this approach, and, as a consequence, the theoretical
physics profession rose rapidly to a place of recognition and importance equal to that enjoyed earlier only by experimental physics. The

rise occurred most rapidly in Germany and in neighboring German-speaking lands, and Einstein, though following a path typical of the course of career advance, rose more rapidly than most to the pinnacle of the new profession.

THE IMPACT OF THEORETICAL PHYSICS

Although considerable historical research has focused on the nature of physics and Einstein's contributions to it during the decades immediately preceding World War I, historians have directed only a little attention to the rise of professional disciplines such as theoretical physics. Christa Jungnickel and Russell McCormmach, authors of the most extensive account, suggest that Einstein's career was typical of German-speaking theorists: most had to enter the profession through experimental work or mathematics, and most had to do so outside of Germany.[8]

Even then, Einstein started at a considerable disadvantage. As a Swiss patent clerk, he possessed little teaching experience; he had studied at a technical college, not at a university; and he had few academic connections. But he did have papers on relativity theory, quantum physics, and statistical mechanics, and in the end that made all the difference.

University employment required a doctoral degree, which Einstein finally received from the University of Zurich in 1906. University teaching also required a postdoctoral certificate, the *Habilitation*. This Einstein did not obtain until 1908, when he began lecturing privately at the University of Bern. In 1909 the University of Zurich opened a second physics position, an "extraordinary" position, to be filled by a theorist, in addition to the full professorship occupied by an experimentalist, Alfred Kleiner. Thanks to his long association with Kleiner, Einstein got the job. He quit the patent office, moved his family to Zurich, and began a heavy load of teaching and laboratory assistance.

As Einstein's position at Zurich indicates, theoretical physics was still considered a second-rate profession. But Einstein's new theories, not his teaching, were what most impressed the Zurich officials. As is true to this day, professors are paid to teach, but they are hired, fired, and promoted primarily for their research and publications. When Einstein left Zurich in 1911, he was succeeded in turn by three leading contributors to theoretical physics who were not especially known for their teaching skills: Max von Laue, Peter Debye, and Erwin Schrödinger. The last of these became, in 1921, the first full professor of theoretical physics at the University of Zurich. All four occupants of that position had come from Germany (though not all were German) and eventually returned to prestigious positions in Germany.

In 1911 Einstein left Zurich briefly for the German University in Prague. Again the faculty was looking not for a teacher but for a leader at the forefront of the new relativity and quantum theories. With Franz Kafka and other notables in Prague at that time, the city was at its cultural zenith. But Einstein was unhappy, and so was his wife. The students were indifferent, his classes only sparsely attended, anti-Semitism was on the rise, and his Serbian wife felt uncomfortable amid the nationalistic tensions in the Bohemian capital. About 1½ years later Einstein returned to Zurich, this time as full professor at his alma mater, the ETH. There he taught theoretical physics to advanced teaching candidates—the very subject that he had so missed a little over a decade earlier.

No sooner had Einstein resettled in Zurich, however, than Max Planck and Walther Nernst arrived from Berlin to offer him a top post in the "capital city" of physics at that time. The delegation informed Einstein he would receive a paid research position with the Prussian Academy of Sciences, a professorship without teaching obligations at the University of Berlin, and directorship of the planned state-sponsored Kaiser Wilhelm Institute for Physics. Now tired of teaching, Einstein would be employed as a theoretical research physicist and expected to perform as such. Teaching and other qualifications and obligations were secondary.

It was an offer that Einstein could hardly refuse. In April 1914 Einstein and his family moved to Berlin, where he remained until Hitler's rise to power in Germany nineteen years later. On July 2, 1914, just one month before the outbreak of World War I, Einstein presented his inaugural address as a member of the Prussian Academy of Sciences, titled "Principles of Theoretical Physics." In it Einstein attempted to reassure those still uncertain about the new discipline with "a few general remarks on the relation of my sphere of activity, which is theoretical physics, toward experimental physics."[9] Einstein the theoretical physicist had reached the pinnacle of his career, and he entered into a period of productivity and controversy that would soon catapult him onto the world stage.

But for Einstein's neglected and unhappy wife, the great man's professional triumph coincided with the end of their personal relationship. Shortly after arriving in Berlin, Mileva separated from her husband and returned to Zurich with the two boys. For years Einstein's success in his work had been drawing him ever farther from his family. In 1913 Mileva wrote to a friend, "My great Albert has in the meantime become a famous physicist. . . . He works tirelessly on his problems; one can even say that he lives only for them."[10] Her own scientific career having failed, and probably never quite over the loss of her first child, Mileva grew more despondent as the years passed. Einstein reacted rather callously with even greater detachment from her, which apparently intensified her feelings. It has been revealed recently that even before Einstein moved to Berlin he had begun a secret, long-distance romance with his divorced first cousin, Elsa Einstein-Löwenthal, who was in Berlin.[11] The romance blossomed in Berlin after Mileva's departure. Einstein finalized the divorce from Mileva in 1919, whereby she received custody of the boys and a promise of the Nobel Prize money, should Einstein receive the prize (he did so in 1922, for the year 1921). Within months of the divorce Einstein married the motherly Elsa, who had two grown daughters from her previous marriage. (She later accompanied him to the United States, where she died in 1936.)

For Einstein the pattern was set. Physics and the search for the secrets of "the Old One," as he called it, would come before all else, including wife, family, and even himself. Moreover, it was clear that he would not let the attraction that his fame and charm held for women go unrequited over the years. Ironically, the man who soon displayed such public concern for human dignity was in fact remarkably limited in his need and ability to relate to those closest to him. Einstein himself declared as much in 1930: "My passionate sense of social justice and social responsibility has always contrasted oddly with my pronounced lack of need for direct contact with other human beings and human communities. I am truly a 'lone traveler' and have never belonged to my country, my home, my friends, and even my immediate family with my whole heart; in the face of all these ties, I have never lost a sense of distance and a need for solitude—feelings which increase with the years."[12]

PHYSICS AND MATHEMATICS

One of the leading factors both in the reception of relativity theory in Germany and in its further development was the relationship between mathematics and theoretical physics. One of the leading centers in the world for mathematics and mathematical physics was the northern German university town of Göttingen. In the throes of industrial revolution, mathematics was under siege in Germany for its seeming abstraction and lack of practical application. Mathematicians, notably Felix Klein at Göttingen, hastened to answer this critique by applying their discipline to more tangible subjects, such as physics. In so doing, they revived an old notion that went back at least to the time of Gottfried Wilhelm Leibniz (1646–1716): the notion that there exists in nature a metaphysical "preestablished harmony" between mathematics and physical reality, perhaps associated for some with the existence of a rational intelligence in nature. For an unknown reason, nature behaves in accord with the rules of mathematics, a human invention. Because of this, mathematical harmonies among concepts in the metaphysical realm were held to correspond to relationships among objects in the real world. Rather than relying only upon principles "wormed out of nature," as Einstein put it, through the study of assemblages of empirical data, believers in preestablished harmony argued that they could discover new laws of nature simply by pursuing mathematical harmonies and manipulating the equations that result from them—secure in the belief that such abstract work corresponded in the end with reality. This belief exerted a profound impact upon the reception of relativity in Germany and, eventually, upon the contemporary conception of theoretical physics. Both of these effects can be traced back to Göttingen mathematics.[13]

Göttingen mathematicians engaged early in the detailed study of Lorentz's electron theory. In a chapter appropriately titled "Physics in the Shadow of Mathematics," Lewis Pyenson has examined a seminar at Göttingen in 1904 devoted to the electron theory.[14] He asks why its members failed to arrive at relativity theory or even to obtain the Lorentz transformations. The answer, says Pyenson, lies in their approach: smitten with the above beliefs about mathematics, they bothered too much with the mathematics and not enough with the physics. In other words, they hoped to resolve the difficulties in electron theory through mathematical manipulations and extensions, not through new conceptual principles rendered consistent with observable phenomena as Einstein did.

In Göttingen Hermann Minkowski formulated a concise mathematical representation of Einstein's theory of special relativity by introducing four dimensions (three of space and one of time), instead of the usual three-dimensional space, to describe the positions and motions of objects such as speeding electrons. Many saw Minkowski's new and more concise mathematical language, not Einstein's conceptual arguments, as the proper formulation of the theory. As a result, the theory as such spread rapidly among German scientific circles.

Yet many of those who grasped at Minkowski's mathematics betrayed their misapprehension of Einstein's work. Some could not comprehend relativity theory as a theory of measurement; others wanted to interpret Minkowski's four-dimensional space-time as a vehicle through which to reintroduce the ether. Many regarded Minkowski as having demonstrated the existence of a preestablished harmony, with the upshot that for them pure mathematics would now serve as the primary tool of research in theoretical physics. They insisted even more that mathematical manipulations of elegant formalisms would yield new insights into nature's laws.

Einstein, of course, could not have disagreed more with the purely mathematical approach, although he did offer a paean to preestablished harmony in 1918.[15] For Einstein, theoretical physics was more than mere mathematics. Mathematics, like experimental data, was a tool, or a set of clues, for gaining insights into nature's laws, but the resulting theory would be more than either equations or data: it would be a new physics, a new comprehension of nature's workings, preferably a new physical worldview far exceeding the methods of its construction.[16]

The contrast can be observed between Minkowski's work in 1908 and that of Einstein and his assistant Jakob Laub at the same time. Using his new formalism, Minkowski presented an elegant new formulation of the relativistic equations for the electromagnetic field. In contrast, Einstein and Laub made every effort to relieve the mathematical burden on their readers. "Because of the circumstance that [Minkowski's] paper places rather high demands on the reader in a mathematical sense," Einstein and Laub declared, they rederived his equations "in an elementary way." In the second portion of their paper, titled "On the Fundamental Electromagnetic Equations for Moving Bodies," Einstein and Laub proceeded beyond the mere equations to suggest an actual experimental test.[17]

Following Minkowski's unfortunate early death in 1909, others followed in his footsteps in Göttingen. Among the most famous and influential were David Hilbert and Hermann Weyl. Hilbert, in particular, undertook a complete program to infuse physics with mathematical rigor. He is even said to have declared at one point, "Physics is too important to be left to physicists." By the outbreak of World War I, the ideology of preestablished harmony and the efforts of Hilbert, Weyl, Klein, and others to pursue it in the equations of physics attracted numerous younger physicists to the mathematical approach, many of whom became leaders of the next generation of theorists. These same young people tended to regard the more conceptual "classical," physical, aesthetic, worldview approach of Einstein and Planck as old-fashioned.[18] As a result of this, after the war, younger theorists felt far more comfortable with theories developed largely through abstract mathematics, such as quantum mechanics.

Many of the young men who formulated these theories had strong mathematical backgrounds, and two of their leading mentors, Arnold Sommerfeld and Max Born, had been schooled in the Göttingen tradition. In 1925 Born, Werner Heisenberg, and Pascual Jordan, working in Göttingen, formulated a heavily mathematical version of quantum mechanics known as matrix mechanics.

But, as we shall see, older physicists such as Einstein and Planck, still loyal to worldviews and conceptual approaches, could not accept the new quantum physics, despite its internal logic and its mathematical coherence. Nor could an aroused yet bewildered public easily accept that they were excluded from fully participating in the knowledge of nature uncovered by these abstract theoreticians.

EINSTEIN, MATHEMATICS, AND GENERAL RELATIVITY

Einstein was no stranger to mathematics, nor even to Minkowski. He had taken several math courses from Minkowski when Minkowski taught at the ETH. But, regarding mathematics as a mere tool, Einstein had cut most of those classes in order to explore the riches of theoretical work on his own. Fortunately, at exam time he could rely upon the careful notes of a fellow student, future mathematician Marcel Grossmann.

To Einstein's great dismay, his own physics grew ever more mathematically abstract. It began with the problem of the light quantum and the electric charge. The former was an indivisible unit of light energy, the second an indivisible unit of charge. Both, he believed by 1909, would find their explanation in a new electron theory of continuous fields propagating through empty space.[19] But it is a difficult mathematical task to amalgamate infinitesimal mathematical points (electrons and light quanta) with continuous variables, such as those associated with electric and magnetic fields.

In addition, for Einstein there were at least two other fundamental dualisms that required resolution. One was the simultaneous behavior of light as waves and particles; the other was the duality of electromagnetic and mechanical worldviews. In addition, by 1909 Einstein apparently believed that the unification of worldviews would occur only through a wholly new theory of fields incorporating the relativity principle. This theory would entail the use of mathematically complicated nonlinear differential equations. Since Einstein had already discovered that electromagnetic energy can carry mass through empty space, he believed that the unification of mechanics and electrodynamics would occur on the basis of a new "unified field theory," somehow combining the equations for gravitational and electromagnetic fields, all subject to the relativity principle. This belief lay at the base of his research program in the period between 1909 and 1911, and the search for a unified field theory became Einstein's lifelong, though ultimately unrealized, quest.[20] It is a quest that continues to this day. It has demanded of Einstein, and of everyone since, uncommon mathematical sophistication—a sophistication that by World War I could hardly have discouraged young people from regarding the mathematics as the embodiment of the physics.

Einstein's new gravitational theory—which later also exerted a profound impact on the public—arose, as one might expect, from a new principle, the principle of equivalence. The equivalent quantities were two types of mass: the mass that is attracted to other mass by gravity (gravitational mass), and the mass that resists forces and accelerations (inertial mass). Starting with Newton two centuries earlier, everyone had tacitly assumed that these two masses are equivalent. Einstein was the first to recognize that this is an assumption that, when fully recognized, has profound implications.

Einstein saw that the principle that gravitational and inertial mass are indeed equivalent could be used to deduce a new theory, a "theory of *general* relativity."[21] Special relativity concerned only cases of uniform velocity—motion without change of speed or direction. General relativity, by considering the effects of gravity to be equivalent to those of acceleration, extended relativity to all forms of accelerated motion. The startling result for everyone was that Einstein was able to supersede Newton's puzzling gravitational force, which seemed to act mysteriously "at a distance." He discovered that he could simply replace gravitation by the curvature of space near a massive body. Instead of being attracted somehow to a massive object, a piece of mass near another piece of mass could be seen as following the curvature of space caused by the presence of the attracting object, bringing the two closer together. The curved space was not just our usual, everyday space but the four-dimensional space-time of Minkowski. No wonder physicists became so enamored at the time of Minkowski and of his mathematical representations of relativity in space-time!

The way in which Einstein employed his new principle of equivalence can be demonstrated by an (updated) thought experiment later used by Einstein himself.[22] It is drawn from the experience of those we now call astronauts. Suppose you sit in a space capsule on a rocket at Cape Canaveral with the doors closed just before takeoff. You feel your normal weight. The rocket then takes off; you feel heavier, then finally weightless as your capsule goes into orbit. However, the situation on the ground before takeoff can be exactly duplicated in space if, while you sit in the capsule, the capsule accelerates upward at exactly the acceleration of gravity back on the earth. You would be pushed down into your seat with a force exactly equivalent to your weight, and all the laws of physics would appear precisely equivalent to what they are at rest back on the earth (see fig. 2).

The equivalence must hold also for a laser light beam shot across the top of the rocket's cabin from one wall to the opposite wall. If the rocket is accelerating upward in space, in the time it takes for the beam to move across the cabin to the opposite wall, the cabin will have accelerated upward, and the beam will appear to curve across the cabin and hit below the spot directly across from where it started. To you, riding in the space capsule, the beam appears to follow a curve in crossing the cabin. But this must hold also for the equivalent case of the rocket at rest before takeoff in the gravitational field of the earth; that is, the light beam must also curve across the top of the cabin. In other words, the light beam must act as if it is pulled downward by gravity. Alternatively, the space-time through which it moves can be considered to be curved by the presence of the massive earth. The gravitational field around an object like a star,

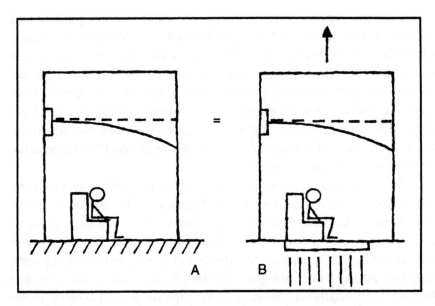

Figure 2. Equivalence and the bending of light. A is in a rocket on the ground before takeoff. B is in a rocket far from any net gravitational pull in outer space. When B's rocket accelerates upward at the acceleration of gravity on the earth, B is pushed down in his seat with a force equal to his weight on the earth. These two situations are completely equivalent to each other; that is, there is no physical means for A or B to tell which situation he is in, without looking outside.

and the attractive force that it exerts on objects, could be replaced by the curvature of space itself such that all objects, such as light or planets or spaceships, follow the curvature, exactly duplicating the motion otherwise caused by the pull of gravity. The more massive the object, the more curved the space-time.

This was certainly a new and radical approach to the old problem of gravity. To formulate a general theory of relativity, Einstein had to find the proper equations for replacing the effect of a gravitational field by the equivalent curvature of space-time. It was an enormous mathematical task, requiring such exotic objects as metric tensors and differential geometry. Fortunately, Einstein's old friend Marcel Grossmann, who had helped him through Minkowski's courses, was professor of mathematics at the ETH while Einstein was professor of physics. Grossmann again came to his rescue; together they explored the mathematical complexities, some of which had been covered in the courses Einstein had cut. Complaints from Einstein abounded in this period. To Sommerfeld he lamented in 1912, "I am now occupying myself exclusively with the problem of gravitation and believe that, with the aid of a local mathematician who is a friend of mine, I will now be able to master all difficulties. But one thing is certain that in all my life I have never struggled as hard and that I have never been infused with greater respect for mathematics the subtler parts of which, in my simple-mindedness, I had considered pure luxury until now![23]

In a series of fundamental papers published in the years 1907–16, Einstein, alone and sometimes with Grossmann as coauthor, presented what he later regarded as his greatest achievement: the theory of general relativity. "You have absolutely no idea," Einstein wrote to a colleague in 1915, "what I went through as a mathematical ignoramus until I arrived in this harbor."[24]

The new space-time abandoned the familiar axioms of geometry that had been laid down by the ancient Greek Euclid and had been taken for granted by almost everyone since. Euclid's geometry corresponded to the infinite, flat space that appears to occur in our everyday world. Since Einstein's new space-time was curved, its geometry had to be non-Euclidean. Einstein utilized the non-Euclidean geometry invented a century earlier purely as an exercise in abstraction by German mathematician Bernhard Riemann.

FURTHER DEVELOPMENTS

Einstein's theory of general relativity, completed at the height of World War I, stood as a monument to human ingenuity. The theory has served ever since as the basis for cosmology, the study of the structure and evolution of the universe. Einstein, Dutch physicist Willem de Sitter, and others were among the first to apply the theory beginning in 1917 to the structure of the universe. When they did so, they discovered that the equations predicted that stars would not be stationary in space but might either collapse together under the attraction of gravity or expand outward into an unbounded universe. Yet the known universe consisted at that time only of our Milky Way galaxy, existing without expansion or contraction in an infinite void. In order to bring his equations into line with such a static universe, Einstein introduced an extra term into his equations multiplied by a constant, now called the "cosmological constant."[25]

The new equations gave the desired result—a static universe. But by 1929 Einstein was calling the cosmological constant his biggest mistake. In 1924 Edwin Hubble, working at the Mt. Wilson Observatory in California, established the existence of the first extragalactic object, the Andromeda galaxy, which lies about one million light-years from the earth. We now know that Andromeda is only one of many billions of galaxies scattered throughout the universe. Then in 1929 Hubble's analysis of objects beyond the Milky Way led him to conclude that the universe is not static after all but is in fact expanding outward. The galaxies of the universe seemed to be receding from each other as if they were expanding from a huge, primordial explosion—the so-called big bang, an event whose existence is since supported by considerable evidence, including the detection of the radiation presumed to be left over from it.

As a result of Hubble's work, Einstein's mistaken constant was no longer needed—or was it? Independent analyses of the expanding universe during the last decade have led to a startling conclusion: the universe is not only expanding, but the expansion is *accelerating*! Some sort of repulsive force or energy seems to be pushing the universe outward at a much stronger rate than the attractive pull of gravity inward. Cosmologists are calling this repulsive force or energy "dark energy," because its origin is unknown. However, it appears so far that its action can be described mathematically by reviving Einstein's old "mistake," the cosmological constant. Research is continuing today to measure the size of this constant and to determine its origin.[26]

The last decade or so has also seen another confirmation of an exotic prediction arising from Einstein's general relativity. A star is a massive object formed by gravitational attraction. It exists for most of its life in equilibrium between the gravitational force pulling its mass inward and the explosive outward pressure arising from thermonuclear fusion reactions within the star. Once the nuclear fuel that powers the fusion reactions begins to run out, gravity suddenly gains the upper hand. The star collapses in a spectacular explosion often observed from earth as a nova or a bright supernova. During the 1930s J. Robert Oppenheimer, the future father of the atomic bomb, and his coworkers were examining the collapse of massive stars using Einstein's general-relativistic equations. They found that if the mass is sufficiently large, the star will collapse into a tiny ball. Then, in 1939 Oppenheimer and his assistants discovered that, according to Einstein's equations, a star of even greater mass could theoretically continue to collapse indefinitely into an infinitely small space, producing such a curvature of space and time that nothing at all, not even light, could escape from the object.

Oppenheimer's strange results were set aside as a mathematical oddity until the discovery during the 1960s of pulsars, stellar objects emitting high-frequency pulses of energy, which were eventually identified as neutron stars. Then in 1967, John Archibald Wheeler called attention to Oppenheimer's earlier prediction of the ultimate state of a massive collapsed star, which Wheeler called a "black hole." But black holes, by their very nature, are difficult to identify directly. Only in the last decade have predictions concerning what should occur in the vicinity of a black hole been confirmed through observations with the Hubble Space Telescope and orbiting detectors of nonvisible radiation. Observations have gone so far in the last few years as to suggest that so-called supermassive black holes, consisting of millions of absorbed stars, exist at the centers of many galaxies, including our own.[27]

Aside from these exotic consequences of general relativity theory, the theory has recently enabled a very practical application— the Global Positioning System. Originally established for military positioning and missile guidance, the system consists of twenty-seven satellites placed in stationary orbits around the earth. When three of the satellites are in view, they can determine the position of a signal sent from the ground by triangulation. The signal travels at the speed of light and by noting the very slight time differences among the receptions of the signal by different satellites, the distance between each satellite and the signal source can be calculated, and the precise location of the signal can then be determined by triangulation. A timing signal from a fourth satellite is used in order to provide a precise location for each satellite and a baseline for time measurements. However, if the positioning is to be completely accurate, all of the measurements of time and distance must also take into account the curvature of space-time in the vicinity of the massive earth, according to Einstein's theory of general relativity. Without general relativity, the accuracy of the entire system would be so faulty as to render the entire exercise useless.

Einstein himself utilized general relativity as the basis for his lifelong search for an even more general relativistic field theory—a unification of gravitation and electromagnetic fields into a single theory. Many physicists today are equally driven by the search for a so-called unified field theory, and they have had considerable success. Four distinct forces of nature are recognized at present: gravitation, electromagnetism, and two nuclear forces. Although distinct in their behavior, they are thought to have been once united at the moment of the big bang, the origin of the expanding universe as we know it. Re-creation of the high energies involved in the big bang, even for tiny fractions of a second in big accelerators (which utilize special relativity theory) can yield new evidence about the nature of the unification to be achieved. A proposed unification of electromagnetism with the two subatomic forces has been confirmed in the last few years with the discovery of all of the elementary particles, called quarks, predicted by the theory. But the unification of this theory with Einstein's gravitational theory is still at the forefront of research, with many exotic possibilities under exploration.

At the same time, scientists' perception of the aims and methods of attaining a unified theory have altered somewhat since Einstein's day. Although scientists share Einstein's aesthetic notions of unity and simplicity, a modern-day unified field theory is foremost a quantum theory; the quantum is not to be explained but is to be built into a theory that also encompasses the continuum of relativistic gravitational fields. Mathematical symmetries and harmonies are regarded not just as aesthetic qualities in Einstein's sense but as indications of physical symmetries; and abstract and speculative mathematical manipulations of the formalism are not only possible but required in order to gain new insights into the physics. Aesthetics, harmonies, mathematical formalism, and experimental data constitute today the synthesis that is contemporary theoretical physics. It is a synthesis that differs considerably from that envisioned by Einstein and the theoretical physics of his day, but it is a synthesis that would have been unthinkable without him.

IMPACT
ON
PHYSICS

| THE QUANTUM |

During his "miraculous year" of 1905 Einstein published not only his theory of special relativity and a strong argument for the existence of atoms in his explanation of Brownian motion but also the first of his papers on the existence of the energy quantum. As discussed in chapter 3, scientists had assembled overwhelming evidence by 1905 indicating that light and other forms of electromagnetic radiation are propagated as waves. Einstein, to the contrary, hypothesized that under certain circumstances light in fact behaves like tiny particles—packets, or "quanta"—of energy. Soon he found that even the motions of atoms in a solid lattice exhibit a quantum behavior.

Whereas Einstein's work on relativity theory and Brownian motion arose as extensions of late nineteenth-century mechanics and electrodynamics, his contributions to quantum theory represented a radical departure from the physics of his day. Einstein himself called his hypothesis of light quanta "very revolutionary"; his discovery of the quantum behavior of matter made a radical revision of Newtonian mechanics inevitable at the atomic scale.[1]

All of Einstein's early works exerted profound impacts upon the growth and nature of physics, especially theoretical physics. The impact of the theories of special and general relativity were the most immediate in this regard. The quantum, in contrast, required decades to make itself fully felt. Ironically, to the end of his life, Einstein often stood apart from most of his colleagues regarding the meaning and uses of the quantum. As Abraham Pais has pointed out, from 1905 until 1923 Einstein was practically alone in taking seriously the existence of the light quantum. From 1926 to the end of his life, Einstein led the opposition to the new quantum mechanics and the paradoxical and indeterministic interpretation of it formulated by younger physicists. While the mainstream of physics flowed toward the quantum mechanics of atoms and elementary particles after 1927, Einstein continued his lifelong quest for a comprehension of the very existence of the quantum in a unified theory of gravitational and electromagnetic fields.[2]

LIGHT QUANTA

Einstein had actually presented two quantum arguments before 1910: the first, his "heuristic" suggestion that light behaves as if it consists of light quanta, or photons; the second, a series of papers on the quantum nature of the energies of particles embedded in matter. Although he convinced himself that these different forms of energy quanta really do exist, he was as puzzled as anyone about what to do with them. He regarded the appearance of quanta in current theories only as an indication that a completely new theory, encompassing both continuous energies and discontinuous energy quanta, was needed.

Einstein's argument for light quanta was well founded. Again, as noted in chapter 3, worldviews and statistical mechanics led Einstein to hypothesize their existence; he could readily account for such puzzling experimental observations as the photoelectric effect using light quanta; and he showed that Planck's energy distribution formula for black-body radiation could be consistently derived only on the assumption that the energies of both the radiation and the particles emitting the radiation are quantized.

Such arguments failed to convince most physicists, however. Continuous waves, not discontinuous packets of energy, still provided the only explanation for many observed features of light, such as the appearance of rays diffracted through tiny slits. Only German experimentalist Johannes Stark rose to the defense of light quanta, much to the annoyance of his colleagues. Even Max Planck was skeptical, despite his strong early support of relativity and Einstein's derivation of Planck's own equation. In the recommendation of Einstein for the position at the Prussian Academy, Planck joined other leading Berlin physicists when he wrote, "That he may sometimes have missed the target in his speculations, as for example in his theory of light quanta, cannot really be held against him. For in the most exact of natural sciences every innovation entails risk."[3]

Newton had actually argued centuries earlier that light is a stream of individual particles, not waves. But by the time of Einstein, nearly a century of experimental and theoretical research on light had led to the unanimous conclusion that light must consist of waves. When in 1915 US physicist Robert A. Millikan confirmed Einstein's equation describing the photoelectric effect, he could hardly bring himself to accept the quantum hypothesis on which the equation rested: "Yet the semicorpuscular theory by which Einstein arrived at his equation seems at present wholly untenable," he wrote.[4]

In 1923 another US experimentalist, A. H. Compton, confirmed that when a beam of light hits a free electron, the light and electron bounce away from each other like two billiard balls colliding. They do so precisely as predicted from the hypothesis that light consists of little particles carrying energy and momentum. To Sommerfeld, Compton's confirmation "tolled the death knell of the wave theory of light." Einstein had made the problem of light quanta even more difficult for his colleagues than a mere choice between particles and waves; he had introduced what was later called the "wave-particle duality." The wave theory of light could not be abandoned, since it worked so well; thus it must be supplemented somehow by the equally valid yet contradictory view of light as particles. In 1909 Einstein demonstrated that the random fluctuations of the energy and momentum of a tiny mirror suspended in a box filled with light resulted in an equation containing the summation of two terms. One term arose from the wave nature of light, the other from its particle nature. Both, he showed, were indispensable for explaining the behavior of light, but neither was compatible with the other. The problem that Einstein posed for physicists was not to choose particles over waves but to combine *both* the wave and particle behavior into one conception of light. "The next phase of the development of theoretical physics," Einstein told the Salzburg meeting in 1909, "will bring us a theory of light that can be understood as a kind of fusion of the wave and emission [particle] theories."[5]

Einstein believed that the fusion would arise from his search for a unified field theory, just as would the fusion of electric charges and the electric fields they generate. But history took a different turn. In an effort to find a proper interpretation of the new equations of quantum mechanics, in 1927 Niels Bohr in Copenhagen incorporated the wave-particle duality into physics without actually resolving it. In his so-called principle of complementarity, certain notions such as waves and particles, though mutually exclusive, are jointly necessary for obtaining a complete picture of nature on the atomic scale. Light waves and particles are "complementary" to each other—jointly essential yet mutually exclusive.

When performing their experiments, physicists make a choice between observing light as a wave or as a particle, but in so doing they disturb nature and destroy the possibility of gaining a complete understanding of nature. In other words, the experimenter becomes part of the experiment, and nature can never be observed independently "as it really is." These and other notions, built into what became known as the Copenhagen Interpretation, gained wide acceptance among most physicists—primarily because of their practical value—but not with Einstein.

QUANTUM SOLIDS

In contrast to the skepticism that greeted the idea of the light quantum, numerous physicists rallied to the quantum theory of matter. Here there were no contradictions with long-cherished notions, except for the already suspect Newtonian mechanics at the atomic level. There was strong experimental support for Einstein's account of an important phenomenon: the behavior with changing temperature of the specific heat of a solid, that is, the amount of heat required to change its temperature. Heat is the random motion of atoms. Regarding the atoms of a solid as oscillating in a lattice, Einstein showed in 1907 that the specific heat should decrease with decreasing temperature—but only if the oscillations are quantized. Newtonian mechanics yielded only constant specific heat.[6]

Walther Nernst's experimental confirmation in 1910 of Einstein's theoretical predictions set off a wave of interest in the quantum theory of matter that reached tidal proportions in the wake of the first Solvay Congress in 1911. This was an international meeting of leading theorists and experimentalists called together by Nernst, with the financial support of Ernst Solvay, a Belgian philanthropist. The purpose was to examine problems with current theories of matter, to study the quantum hypothesis, and to begin the search for "a future mechanics." Einstein and others addressed the congress, held in Brussels, and through the addresses and discussions, most physicists gained their first full acquaintance with the problems of quantum theory and the current state of research.

Most participants in the 1911 Solvay Congress left with the sense that they were standing on the threshold of a new era in physics. Jungnickel and McCormmach observe a sharp increase of publications on quantum theory after 1911. After the publication of the proceedings in French, many of these publications occurred outside of Germany.[7] Einstein, meanwhile, returned to Prague to contemplate quantum physics (with little success) and relativity theory (with great success).

Einstein and the physics community parted company at this point regarding the quantum theory of matter. While Einstein tried to embed the discontinuous light quanta in the continuous electromagnetic field and otherwise struggled with general relativity, others turned with enormous success to the atom. Niels Bohr, at that time a Danish postdoctoral researcher in the Cambridge laboratory of Ernest Rutherford, learned of the latest quantum theories upon Rutherford's return from the Solvay Congress. He soon applied the notion to Rutherford's problematic model of the atom. Rutherford pictured atoms as consisting of electrons orbiting a nucleus much as planets orbit the sun. Assuming quantized energy and the emission and absorption of light quanta by the atom, Bohr managed to derive important observed properties of the hydrogen atom, notably the appearance of certain spectroscopic lines emitted by the atom when heated.

The Bohr model of the atom became the basis of a full-fledged quantum physics developed in Germany during World War I by Arnold Sommerfeld and coworkers. At last converted to relativity and the quantum, Sommerfeld joined the two theories into a relativistic quantum theory of the atom. In a series of brilliant papers, Sommerfeld set Bohr's model on surer mathematical footing and extended it to include relativistic mass changes of the orbiting electrons moving in three dimensions. His account of hitherto inexplicable spectroscopic effects rendered Sommerfeld one of the world's leading atomic theorists. The new Bohr-Sommerfeld quantum theory of the atom set the stage for German world leadership in atomic physics after the war. Sommerfeld's widely used textbook on the subject and his renowned institute in Munich made him for years one of the foremost teachers and mentors of the new postwar generation that produced quantum mechanics.

GERMANY AND THE ATOM

Although Germany could boast of such luminaries as Einstein, Planck, and Sommerfeld, a strong tradition in theoretical physics, and the most established theoretical physics profession in the world, the country after World War I seemed an unlikely setting for world leadership in atomic physics. While the international community subjected Germany to academic boycott, internally the defeated nation teetered near economic and political collapse. A raging inflation reached astronomical proportions by 1923 in concert with political violence and revolution on the left and right, culminating in Hitler's unsuccessful Beer Hall Putsch in Munich in 1923. While their elders seemed bewildered and mistrustful of the new state of affairs, young people reacted with anger and resentment, and universities became hotbeds of extremism.[8]

Why did quantum atomic physics thrive in Germany during those turbulent years? Of course, physics and allied sciences were not the only cultural endeavors to thrive: Weimar art, literature, music, and theater also reached extraordinary new heights. Several historians, most notably Paul Forman, who have examined the financial and ideological dimensions to Weimar physics, have found that the growth of German atomic physics owed much to the prompt and forthright measures undertaken by the older generation. Recognizing theoretical atomic physics as one arena in which Germany still could best her international competitors, older theorists astutely gathered funding for their programs from well-heeled German industrialists as well as from US philanthropists dismayed at the treatment of Germany and eager to influence German culture.[9]

The financial support flowed into a new foundation, the Emergency Association of German Science. In an innovative departure from the usual practice of simply handing out research funds, the new foundation invented the modern project-grant system of support, first employed in the United States only after World War II. Researchers submitted grant applications that were subjected to peer review by independent specialist committees that evaluated the proposed projects and awarded funding according to merit and desirability of the research. Thanks to Einstein, Planck, and other Berlin scientists, the Emergency Association funneled the largest share of support into physics. The physics disciplines most preferred were atomic physics, radiation, and the structure of matter—in particular, theoretical and experimental research on the relativity and quantum theories.

In addition to financially preferred areas of project grant funding, the Emergency Association created another novelty, now also commonplace: stipends for promising postdoctoral researchers who, because of the inflation, would otherwise have had to leave physics. Thanks again to Planck, Einstein, and others, many of the early stipends went to young people such as Werner Heisenberg, who would help to create the new quantum mechanics.

A third innovation arose from the US-based Rockefeller Foundation. In 1923 its International Education Board established traveling fellowships for young foreign postdocs, especially physicists, accompanied by the generous funding of new research facilities and equipment in financially strapped institutes. Bohr and Einstein, both of whom found the boycott of German science repugnant, were instrumental in gaining Rockefeller funding for their German colleagues. Since many of the younger physicists had gravitated to atomic physics, Bohr's Copenhagen institute, which itself received generous Rockefeller construction funds, became a magnet for traveling Rockefeller fellows.

The availability of research grants and stipends and the close connections between Bohr and German atomic physicists resulted by the early 1920s in a series of modern research institutes, of which three formed a "quantum triangle" trodden by nearly every young German atomic theorist. Each displayed its own characteristics. With Sommerfeld in Munich they learned spectroscopy and atomic theory; with Max Born in Göttingen they learned sophisticated methods of atomic model calculations; and with Bohr in Copenhagen they learned "philosophy," the intensive study of the conceptual foundations of quantum atomic physics. Berlin, though still the "capital of German physics," remained in this period a stronghold of the older generation—Einstein, Planck, von Laue, Nernst, and Fritz Haber—who preferred the more "classical," prewar mode of research: the search for physical worldviews that would unite all of physics and provide a deeper insight into the universal structure of the physical world.

Not all travelers on the quantum triangle were German. Italy sent Enrico Fermi; the Netherlands sent S. A. Goudsmit; and the United States sent everyone it could. The United States was also the most diligent in inviting all three mentors—Born, Sommerfeld, and

Bohr—to come and lecture, while sending its brightest young theorists to learn the new physics at its source (all despite the boycott). One visitor to Born's institute remarked: "In the winter of 1926 I found more than twenty Americans in Göttingen, at this fount of quantum wisdom."[10]

A brief overview of the backgrounds of the surprisingly few main contributors to quantum mechanics reveals the success of the Planck-Einstein strategy of funneling financial support to young people in atomic physics. By 1926 two forms of quantum mechanics had been achieved, primarily by German-speaking physicists: one was the so-called matrix mechanics, the highly mathematical physics created primarily in Göttingen; the other was the more mathematically accessible wave mechanics, arising from the work of Erwin Schrödinger, Louis de Broglie, Einstein, and Planck. All these scientists were men, and most stemmed from educated, upper-middle-class families, but there were numerous contrasts between the physicists who followed these two forms of quantum mechanics. The matrix mechanicians worked mainly on the "quantum triangle," researched spectroscopy, published in the new *Zeitschrift für Physik* (Journal for Physics), and, leaving out their mentors, had an average age in 1925 of twenty-four years. The wave mechanicians published mainly in the established *Annalen der Physik*, worked off the triangle on statistical mechanics, and had an average age in 1925 of forty-six years (see also table 1). By 1928 the members of the matrix mechanics group would come to dominate quantum studies and the whole physics profession, after their formulation of the Copenhagen Interpretation.

QUANTUM MECHANICS

The early development of quantum mechanics and the wide success of the theory after 1927 helped to ensure theoretical physics an eminent position within the physics profession. Einstein and his generation played a diminishing role in this development. But Einstein did keep his hand in. In 1917 he showed that the jump of an elec-

TABLE 1   Backgrounds of Some Leading Quantum Physicists

NAME	YEAR OF DOCTORATE	PLACE OF DOCTORATE	AGE IN 1925	PLACE OF WORK IN 1925
FOUNDERS OF WAVE MECHANICS				
L. de Broglie	1924	Paris	33	Paris
A. Einstein	1906	Zurich	46	Berlin
M. Planck	1879	Munich	67	Berlin
E. Schrödinger	1910	Vienna	38	Zurich
FOUNDERS OF MATRIX MECHANICS				
N. Bohr	1911	Copenhagen	40	Copenhagen
M. Born	1906	Göttingen	43	Göttingen
P. A. M. Dirac	1926	Cambridge	23	Cambridge
W. Heisenberg	1923	Munich	24	Copenhagen Göttingen
P. Jordan	1924	Göttingen	23	Göttingen
A. Landé	1914	Munich	37	Tübingen
W. Pauli	1921	Munich	25	Hamburg
A. Sommerfeld	1891	Königsberg	57	Munich
J. H Van Vleck	1922	Harvard	26	Minnesota

tron between quantum orbits in Bohr's atomic model is subject to probability laws. Einstein's objections to matrix quantum mechanics during the mid-1920s helped to spawn an alternative. Matrix mechanics, formulated by Heisenberg, Born, and Jordan, was not only highly mathematical, it was also highly abstract and positivistic: it emphasized laboratory observables, not unobservable entities such as electron motions within atoms. Heisenberg and his colleagues overtly rejected the use of visualizable atomic models, such as pictures of orbiting electrons—one of the most important constituents of classical theories.[11] In April 1926 Einstein and the Berlin physicists invited Heisenberg to an afternoon lecture and debate on his new physics. After fielding numerous objections, Heisenberg accompanied Einstein home to his apartment. Einstein attempted to explain to the twenty-four-year-old that positivistic theories are too limited, even if he himself had relied on observational information earlier in formulating special relativity. "In reality the very opposite happens," he told Heisenberg. "It is the theory that decides what we can observe."[12]

Mathematical erudition and the loss of visualization of physical processes in matrix quantum mechanics also bothered the founders of the alternative wave mechanics. Invoking special relativity and symmetry arguments, Louis de Broglie had discovered the possibility that just as electromagnetic waves can behave as particles, so matter can behave as waves. It was the kind of argument Einstein loved. Experimental confirmation, Einstein's support, and his own researches led Erwin Schrödinger to the complete equations of wave mechanics. He consciously viewed this as an alternative to Göttingen's more abstract matrix quantum mechanics. The two turned out to be mathematically equivalent: either way of writing the equations gave the same answers. In a famous footnote to his paper proving this equivalence, Schrödinger declared, "My theory was inspired by L. de Broglie . . . and by short but incomplete remarks by A. Einstein. . . . No genetic relationship whatsoever with Heisenberg is known to me. I knew of his theory, of course, but felt discouraged, not to say repelled, by the methods of transcendental algebra, which appeared difficult to me, and by the lack of visualizability." Einstein wrote Schrödinger that he was convinced his work represented "a decisive step forward . . . just as I am convinced that the Heisenberg-Born approach is off the track."[13]

By 1928 the two formulations—matrix mechanics and wave mechanics—had been joined into one theory, a new "quantum mechanics" to replace the old Newtonian mechanics when dealing with phenomena on the atomic scale. To connect concepts on the scale of everyday phenomena with atomic events ruled by quantum mechanics, Heisenberg, Bohr, Born, and others formulated their Copenhagen Interpretation of the quantum equations. Despite its strange and radical features, it found wide acceptance following the fifth Solvay Congress, at the end of 1927. Like special relativity earlier, measurement formed the essence of the new theory, despite Einstein's objections.

A basic element of the new Copenhagen Interpretation involved Bohr's complementarity principle: the idea that nature encompasses dualities, and observers necessarily choose one side of a duality in making their observations. The choice causes a disturbance of nature, and the result is described by what Heisenberg called his "uncertainty relations": certain basic properties of an object cannot be measured simultaneously with unlimited accuracy. For example, to measure the velocity (or momentum) of a moving electron with extreme precision, one must interfere with it so as to make its position extremely imprecise. If one measures the position precisely, the velocity becomes uncertain. In other words, we cannot know simultaneously the position and the velocity precisely, only what nature and our instruments allow us to know through our limited measurements. (Quantum mechanics suggests that the particle possesses neither of these until they are measured.) This limitation on such measurements as the simultaneous position and velocity of a particle results in a limitation on predictions of where the moving particle will be at a future time. We cannot predict the precise path it will take. This, in turn, undercuts the principle of causality, a basic assumption underlying all of classical physics. We cannot know the parameters of a particle's motion precisely at a given moment, so we cannot predict where the particle will end up later. As with Einstein's jumping electrons between quantum states, we can say only with a certain probability where the particle is likely to be. This probability, at least, can be calculated exactly.[14]

With the new quantum-mechanical formalism and a well-constructed interpretation of it, in 1927 Born and Heisenberg proclaimed the quantum revolution at an end. In a paper delivered to the fifth Solvay Congress in Brussels they declared, "We regard quantum mechanics as a complete theory for which the fundamental physical and mathematical hypotheses are no longer susceptible to modification. . . . The subsequent development of the theory of radiation will change nothing in this state of affairs."[15]

Einstein could not have disagreed more. In a series of papers and in a running debate with Bohr that started at this Solvay Congress and lasted to the end of his life, Einstein raised objection after objection.[16] First of all, he believed, nature must be regarded as independent of the observer and knowable by us as it really is; second, the wave-particle duality must be resolved, not assumed as a given; third, strict causality and determinism of events must be restored, and probabilistic predictions regarding measurements of experimental results thereby removed. Nature is knowable, and our work as scientists is to know it. Contrary to Bohr and Heisenberg, Einstein insisted that since quantum mechanics is probabilistic, it is not yet complete; once completed, it will not be probabilistic. His most famous statement on the probabilistic nature of quantum results is both succinct and profound. It appeared in a letter to Born in 1926: "The quantum mechanics is very worthy of regard. But an inner voice tells me that this is not the true Jacob. The theory yields a lot, but it hardly brings us closer to the secrets of the Old One. In any case I am convinced that *He* doesn't throw dice."[17]

Einstein's scientific opponents, armed with a logically closed and coherent physics, carried the day. Once regarded as too radical in his quantum thinking, Einstein now appeared as too conservative, a holdover from the late nineteenth-century "classical physics" that a new generation had replaced with quantum mechanics. Einstein's objections were consistent with his earlier, essentially nineteenth-century notions, but his argument remained nonetheless radical: the quantum had to be understood, not merely accepted. Einstein was in fact calling for an even more radical revolution than that contemplated in Copenhagen—an entirely new theory that went far beyond quantum mechanics to a complete understanding of the quantum itself. To the end of his life Einstein refrained from accepting the views of his colleagues, as quantum mechanics proceeded without him from one apparent triumph to another in the study of nuclei, elementary particles, solids, and stellar phenomena. Only time will tell if Einstein was right, after all.

The triumph of quantum mechanics over the objections of Einstein (and a few others) owed also to two remarkable institutional factors regarding theoretical physics. During the years following the Solvay Congress in 1927, nearly every quantum mechanician carried the new "Copenhagen spirit" (as they called it) to foreign shores and around the world. The United States proved particularly receptive to the missionaries, and the reception owed greatly to the practical utility of the theory.[18] Quantum mechanics proved itself in wide numbers of applications, as it has ever since. Convinced of the practicality of quantum mechanics, Americans and others decided to leave erudite debates over ultimate reality and knowledge of nature to the "philosophers."

Second, the older generation of German theoreticians was beginning to vacate its teaching chairs. Within a few years nearly the entire band of young men who had created quantum mechanics found themselves in charge of central European theoretical physics institutes. These institutes quickly flooded with excellent students from Germany and abroad who flocked to the new centers to learn the new physics. During the five years between the 1927 Solvay Congress and Hitler's rise to power in Germany, the founders of quantum mechanics trained a new generation wholly within the new physics. Then, as Germany sank into dictatorship, they scattered around the world, armed with outstanding early successes in applying the new physics to a host of fundamental phenomena, from magnetic materials to nuclei to elementary particles. As the new generation dispersed, they found new seed centers for a worldwide profession of quantum mechanics outside of central Europe. The phenomenon was symbolized, ironically, by Einstein's emigration from Berlin to Princeton in the United States, although Einstein himself opposed the new physics, had few students, employed only a handful of assistants, and never founded a school of research.[19]

Despite Einstein's decreased participation in quantum mechanics, the "intellectual mastery of nature" made possible by the rise of theoretical physics through the creation of relativity theory and quantum mechanics set the stage after 1927 for the "material mastery of nature" that ensued during the second half of the twentieth century.[20] The new profession first united quantum physics with special relativity theory, then pushed the new physics into the atomic nucleus and out to the cosmos. Many new and profound practical applications have since sprung from the quantum revolution; among them are nuclear energy and nuclear weapons, lasers, semiconductors and superconductivity, the invention of the transistor—the foundation of the digital computer revolution—the prospect of quantum computers, and the startlingly strange cosmologies of recent years. These are only a few of the many material and intellectual consequences of quantum theory that have transformed our world and made it what it is today.

RELATIVITY REACHES THE PUBLIC

Many modern technologies owe their theoretical origins to Einstein, and the current research and professional standing of the modern physicist would be unthinkable without him. In looking beyond science and technology to the broader society and culture, it is remarkable to find that Einstein's "impact" seems equally pervasive across a wide variety of disciplines and developments—in ways that one would hardly expect from highly technical theories that only specialists can fully assimilate.

Despite C. P. Snow's oft-cited lament over the divide separating the "two cultures" of the sciences and the humanities, Einstein seems to have bridged the gap—at least in some respects. Since Einstein was involved only indirectly in quantum mechanics, this story will focus on relativity theory. We begin with the questions of how and why Einstein's theories of relativity were so influential within the broader culture and society. Were the technical theories themselves influential or only what people *perceived* the theories to mean and to imply? Clearly the public believed that they did imply much more than mere abstract physics; otherwise they would have been totally ignored. This reaction can be traced to the public conception of science in the early twentieth century.

Even more so than today, faith in science to uncover fundamental truths about nature was so strong that many believed that the picture of the world uncovered by fundamental sciences such as physics possessed universal characteristics that went far beyond the narrow confines of abstract theories. Ignoring Einstein's repeated objections, many believed that, because scientific knowledge was so fundamental, the worldview presented by physics could be expanded to a view or philosophical position regarding the entire world, including cultural, ethical, even religious and political affairs. As a result, impressive new scientific discoveries naturally elicited public responses of either affirmation, disapproval, or befuddlement—not for their scientific merit but for their presumed human and social implications. Einstein found such responses not only objectionable but frankly repulsive: applications of science to areas for which it is not applicable is a gross abuse of science and can only lead to demagoguery, he argued. In one of many statements on this issue he declared, "I believe that the present fashion of applying the axioms of physical science to human life is not only a mistake but has something reprehensible to it."[1]

Yet apply physical theories to human life the public did, and in ways that spread the impact of relativity far beyond physics. Einstein scholar Gerald Holton, in an insightful essay entitled "Einstein and the Shaping of Our Imagination," has suggested how Einstein's impact on culture might have occurred.[2] Naturally the impact was quite varied, for the understanding of relativity varied widely, as did ideas on what it might imply about nature and about us. On the negative side, these reactions ranged from fascinated bewilderment, enthusiastic misapplications, or the despair of humanists confronted with tensor algebra, to outright hostility, rejection, and exploitation by political and scientific ideologues.

On the positive side, Holton finds that reactions varied from acceptance of the physics throughout science and technology, to incorporation into philosophy and extension of the unified physical worldview espoused by Einstein, to the stimulation of new, creative work.

Especially significant is Holton's assertion that what unites the most positive and creative responses to relativity theory is not at all the extent to which the individual respondents actually understood the theory. Rather, it is the extent to which they assimilated perceived elements of the theory into a new expression of human experience that resonated within the broader culture. "Quite apart from misuse and oversimplification," writes Holton, fecund scientific ideas such as relativity theory "lend themselves . . . to further adaptation and transformation in the imagination of similarly exalted spirits who live on the other side of disciplinary boundaries." Influential theories such as relativity seem to strike a special chord across a range of ideas, or metaphors embodying those ideas, that are in currency at a given time. This chord sets up a mutual resonance in the minds of kindred spirits in other fields, releasing "in them a fruitful response with an authenticity of its own, far removed from textbook physics."[3]

Holton's metaphor might be stretched across an entire age. New scientific ideas such as relativity can provoke not only a resonance among kindred spirits but also a dissonance among less exalted thinkers who misuse and exploit the new theories for political and personal gain. And these resonance-dissonance reactions can spread beyond the cultural elite to encompass a wide spectrum of opinion among the general public.

At this stage of our understanding the resonance metaphor seems to represent the best account of the ultimate "impact" of Einstein's theories of relativity on areas of society and culture far beyond the specialized confines of science and technology. That impact became noticeable in the years of upheaval in Europe following World War I, when a whole new range of metaphors came into play regarding science, nature, and humanity's relation to both.

END OF THE ENLIGHTENMENT

The shocking spectacle of the world's most advanced cultural and industrial nations engaging in brutal trench warfare, starvation blockades, war propaganda, and technological weapons of mass destruction signaled the end of the Newtonian age of Enlightenment and the advent of its successor, the so-called modern era. World War I shook Europe's faith in human progress through enlightened reason and scientific achievement, replacing it with widespread cynicism and a general sense of lost meaning and direction.

The period following World War I was one of great and sometimes frightening changes: the rise to prominence of urban industrial centers; the spread of democracy to an unwilling Germany; the perceived threats to the European liberal tradition posed by the Bolshevik Revolution in Russia and scattered, bloody Communist uprisings throughout central Europe; and the universal sense that somehow people had suddenly lost control of their destinies. These changes signaled, in the words of Oswald Spengler's famous two-volume essay, *The Decline of the West*. It was a decline blamed upon and marked for many by the predominance of lifeless Newtonian mechanism, "soulless" materialism, deterministic thinking, and working-class (i.e., socialist and democratic) "mass culture."[4]

Amid this depressing state of affairs, relativity theory entered into the postwar era as a reinforcement for some and as a contradiction for others.

On November 6, 1919, a team of British astronomers reported to the Royal Society of London an amazing result. During a recent total eclipse of the sun they had observed that the positions of the stars near the sun appeared to have shifted slightly from their proper positions. The apparent shift arose from the bending of light rays from the stars as the light passed by the sun (see fig. 3). This bending of light rays, they found, occurred in full agreement with Einstein's prediction in his papers on general relativity! Sir Arthur Eddington, the astronomer who made the measurements, celebrated with a parody of Omar Khayyám's *Rubaiyat*. The last verse went as follows:

Oh leave the Wise our measures to collate.
One thing at least is certain, LIGHT has WEIGHT.
One thing is certain, and the rest debate—
Light-rays, when near the sun, DO NOT GO STRAIGHT.[5]

The news from London—coming in the same year that a defeated Germany reluctantly signed the Treaty of Versailles—swept across a war-weary Europe like a tidal wave. Newspapers across the continent extolled the confirmation of what some called "general relativity," by which they meant the relativity of everything in general. A fascinated public clamored for explanations, and Einstein, the Berlin professor, became an overnight celebrity. The professor, who always shunned publicity, was at a loss to explain his sudden popularity. He complained to Max Born that he felt like a man

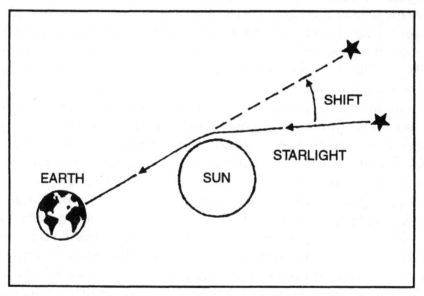

Figure 3. The bending of starlight. In 1919 Einstein's prediction that space is curved near a massive object was confirmed by this experiment. During a total solar eclipse, stars near the sun become visible. If starlight passing near the sun is bent toward the sun owing to the curvature of space, the known position of the star appears to be slightly shifted. This shift occurred exactly as Einstein had predicted.

plagued by the Midas touch—only, instead of turning into gold, everything he touched turned into newsprint![6]

Historians have not been much more successful in explaining the sudden public fascination with Einstein and relativity. Certainly it had much to do with the postwar intellectual and moral upheaval, with the need of many for some sort of prophet to provide answers, and with the new worldview of nature that relativity theory seemed to convey. Historians Alan J. Friedman and Carol C. Donley write: "Einstein proved to be a remarkably rich and authoritative source of idea and metaphor. . . . Einstein's personal image came to represent the power of scientific intellect. Einstein's face was a convenient literal image for discussions of genius, rational thought, and abstract notions."[7]

Long before the war, the public had linked truths about human affairs with truths about nature uncovered by science. Late nineteenth-century mechanical science seemed fully in step with traditional religious and philosophical beliefs. Paralleling moral and religious absolutes, science seemed to be uncovering simple, absolute, visualizable mechanical laws occurring in a world of absolute, objective space, time, and mass.

At the close of the Great War, the familiar features of the mechanical worldview seemed suddenly undercut by relativity theory, and the old view's "deterministic" elements seemed at fault for current uncontrollable difficulties. The assumptions of absolute space, time, and matter suddenly lost their meaning, and strange things began to occur at high speeds. Although the public eagerly sought to learn more, beneath the fascination for relativity lurked a widespread anxiety. With the old order already in shambles about them, many feared that the triumph of relativity theory meant, by its very name, the decline of the old transcendent moral order and the triumph of universal "relativism" in every sphere of life—a gross misunderstanding and erroneous extension of relativity theory. If the basic "given" elements of nature were no longer absolute and independent of human observers, many reasoned, then neither were traditional meanings and beliefs in religion, ethics, and social ideology. Coming on top of the upheavals over Darwinian evolution, science seemed to be telling the public—as it does to some in the United States today—that cherished values and beliefs are a mere illusion.

Different people reacted differently to this somewhat contradictory state of affairs. Many, of course, among them most physicists, hailed the new physics for the great achievement that it was. Others, however, hailed it as a challenge to the deterministic mechanical worldview. One intellectual reaction to the abhorred "mechanism," especially in postwar Germany, is common throughout modern Western history in situations where science seems a threat: a flight from "lifeless" science into romantic, "spiritual," antiscientific modes of thinking. Relativity theory apparently lent new confidence to those who now called for physics to bring forth a new, nonmechanical conception of nature in which fundamental materialism, mechanical notions, and, especially, determinism and causality would no longer hold dominion, and nonmaterial spiritual values and truths could once again gain credence. These were just the scientific features that the creators of quantum mechanics presented to the public in 1927. Paul Forman has argued at length (though not definitively) that these features of the Copenhagen Interpretation arose directly through the "adaptation" of Weimar physicists and mathematicians to the hostile postwar intellectual environment.[8]

Neoromantics and mystics ever since have regarded quantum indeterminism on the microscopic scale as the scientific basis for the existence of *human* indeterminism and free will in the everyday world. Strenuously protesting the mingling of scientific and human truths, Einstein openly rejected both romanticism and quantum acausality.

Although Einstein's theories of relativity also challenged the mechanical worldview, romantic critics never regarded Einstein's theories as the alternative they longed for and demanded from physics. One reason was Einstein's retention of determinism by retaining Newton's laws of motion in inertial reference frames and his defense of determinism against the Copenhagen Interpretation. However, romantics did not miss the opportunity to associate Einstein's relativity theory with ontological relativism—the assertion that all forms of ultimate knowledge, including their own preferences for mystical and religious knowledge, are equally valid accounts of the physical world.

Others, however, simply rejected relativity theory out of hand. The strongest and most vehement early opponents of relativity theory appeared in Germany. Most of the German public at first found cause to celebrate relativity theory. After all, amid military defeat and economic collapse this was a German triumph that even the victorious British could not deny—and relativity theory could be seen as a contradiction to the hated "mechanistic materialism" of the new German social democracy. Yet two other factors were also at work: the perceived challenge to traditional meaning and beliefs and opposition to Einstein himself. In a sense the theories of relativity seemed as incomprehensible to a bewildered public as did the new situation in which they now found themselves. As traditional meanings fell under siege, so did relativity theory. One early opponent of relativity went so far as to label it "degenerate" science, a term later used by the Nazis to stamp out all forms of nonrealistic modernist art and culture.[9]

The vocal opponent was Paul Weyland, the nondescript head of one of the many small proto-Nazi hate groups that sprang from the soil of German upheaval in reaction to the situation in postwar Germany. Many of these groups, such as the "Swastika Majority" in Munich, included in their hatred of democracy and modernity a hatred of Einstein and relativity theory. The focus on Einstein was at first mainly political, but it soon turned increasingly anti-Semitic in the postwar atmosphere.[10]

The outbreak of World War I within months of his arrival in Berlin came as a shock to Einstein, as it did to many other people. The brutal reality of war and the militaristic nationalism displayed on both sides were an affront to his sensibilities and awakened his sense of responsibility as an international citizen of the world. In December 1914 he wrote to a colleague, "The international catastrophe has imposed a heavy burden upon me as an internationalist. In living through this 'great epoch,' it is difficult to reconcile oneself to the fact that one belongs to that idiotic, rotten species which boasts of its freedom of will."[11]

When the overwhelming majority of German professors signed a declaration to the world in support of German militarism, Einstein was one of only three Berlin professors to sign a pacifist, antiwar countermanifesto. As the war dragged on, Einstein became ever more committed to pressing for a democratic constitution in Germany and for the establishment of the international League of Nations to help prevent future wars. After the war ended, Einstein's well-publicized support of the German social democracy (the Weimar Republic) that had replaced the Kaiser and his monarchy provoked the ire of antidemocratic hate groups. But Einstein's ideas never changed. His commitment to social democracy, international understanding, the fundamental dignity of the individual, and the protection of intellectual freedom remained to the end of his life. So, too, did his sense of responsibility as the world's most public scientist to speak out on these issues when necessary. And, as the century unfolded, he found plenty of need to do so.

In August 1920, as the German social democracy took shape, Einstein decided to respond in a Berlin newspaper to the attacks on relativity theory orchestrated by Weyland's Berlin hate group. The response only further incensed Weyland's followers, who drew the support of several active scientists, including an especially prestigious antirelativist, Nobel Prize–winning experimentalist Philipp Lenard. In his 1920 article, Einstein had also criticized Lenard's arguments against relativity theory and called for a scientific debate at an upcoming meeting of the German Physical Society in Bad Nauheim, Germany. Lenard, who saw Einstein's article as an affront, took up the challenge. Weyland and followers prepared for disruption.[12]

Max Planck, the dean of German physics, chaired the tense session on relativity theory. Attempting to avoid an incident, Planck left time only for brief discussion following the scientific papers. Lenard took the floor, and his objections to relativity theory were actually quite reasonable—that is, for a physicist who had failed to comprehend any developments in theoretical physics since 1900. He deplored the "abolition of the ether," objected to the universal validity of the relativity principle, and lamented the loss of visualizability of events. What is visual, Einstein replied, changes over time. A "compromise" proffered by Lenard, entailing limited return of the ether and limited validity of the relativity principle, was rejected out of hand. (A theory is not a matter for negotiation.) With that Planck closed the discussion. Lenard remained unconvinced and consequently became an even more virulent and dangerous opponent of both Einstein and relativity theory.

PHYSICS AND IDEOLOGY

The personal and professional needs of scientists and the cultural needs of the public sometimes coincide. When they do, a resonance often occurs between a scientist's portrayal of a new discovery to the public and the public's reactions regarding its understanding of the natural world. Scientists, like other human beings, are neither above infusing their portrayals of science with personal beliefs nor even above engaging at times in outright manipulation and ideological indoctrination of the public for their own gain. Philipp Lenard represents an especially grotesque and extreme example of this.

In the years following the Bad Nauheim debate, Lenard's opposition to relativity theory became more professionally motivated, anti-Semitic, and ideological. He was soon joined by another Nobel Prize–winning experimentalist, Johannes Stark.[13] Although once the sole supporter of Einstein's light-quantum hypothesis, Stark refused to accept the Bohr-Sommerfeld quantum theory of the atom and, later, quantum mechanics. A combination of pathological personality traits and professional difficulties drove both men to seek a public outlet for their frustrations and a vehicle for personal influence through early support of Hitler's Nazi party.

Many of those who rejected the German Weimar democracy also rejected relativity and quantum theory. Lenard and Stark became actively political when they found themselves in disagreement with the leaders of a now stratified physics profession and with the direction in which that profession was moving. This only reinforced their dissatisfaction with the direction society was taking and encouraged the mixing of their contempt for society with their contempt for the physics of the ruling elite. Some historians have related Stark's scientific position to the professional divide between the German provinces and the dominant Berlin leadership, including Einstein. The latter pursued the "modern" research areas of relativity and quantum theory, while the former clung to the more traditional "classical" fields of technical and industrial physics.[14]

In addition, Lenard and Stark seemed particularly resentful of the sudden rise of the theoretical physics profession to prominence under Einstein and Planck and of the sudden adulation accorded Einstein for work that, following nineteenth-century tradition, Lenard and Stark regarded as less fundamental than their own experimental research. As early as 1912 Lenard objected to the creation of a new associate professorship in theoretical physics at his university, Heidelberg. A letter from a colleague to Einstein's assistant, Jakob Laub, described Lenard's position: "He wanted to deny theoretical physics any right to an independent existence [saying] that he was enough of a theoretician himself, etc. For example, he wanted to refuse on principle [to accept] purely theoretical workers as doctoral candidates."[15]

Two years after the Bad Nauheim meeting of 1920, German scientists and physicians prepared to meet again, this time in Leipzig. Lenard published a "word of warning" in an appendix to his new book on the old ether theory. Relativity theory, he proclaimed, was no theory at all, merely a discredited hypothesis. As one author paraphrased Lenard, relativity theory "was nothing more than the shrill cry of the marketplace, supported predominantly by mathematicians who were retreating into a new form of scholasticism: they sought knowledge in their heads instead of experimenting with their own hands and observing nature with their own eyes."[16] Outside the Leipzig meeting hall in which a session on relativity theory took place, Lenard's and Weyland's followers handed out anti-Semitic handbills railing against Einstein and relativity theory. Inside, Max von Laue defended relativity in Einstein's stead. The assassination of Germany's famed Jewish foreign minister, Walther Rathenau, before the meeting had encouraged Einstein to withdraw temporarily from public life.

Following the Leipzig meeting, Lenard's and Stark's ravings became even more infused with anti-Semitism, and, as Alan Beyerchen noted, their writings during the 1920s exhibited close, but independent, parallels with such Nazi writings of the day as Hitler's *Mein Kampf* and Alfred Rosenberg's Nazi attack on the Enlightenment, *The Myth of the Twentieth Century*.[17] The reservoir of anger and hatred that Hitler and his followers were tapping in the broader society resonated with the personal and professional hatreds and frustrations harbored by these physicists, who then used them to lash out at their perceived opponents through attacks on Einstein and modern theoretical physics.

By exploiting public anti-Semitism and anxiety over relativity, Lenard and Stark discovered that they could easily gain the public recognition and status that they felt their profession had denied them. Within days of Hitler's rise to power in 1933, Stark wrote to Lenard, "At last the time has come when we can bring our conception of science and research into play." Lenard wrote to Hitler himself, proposing that in all questions regarding science and mathematics at German universities Hitler should "obtain my counsel before making a decision."[18]

In order to exploit the turn of events to their own advantage, Stark and Lenard cast an ideological sword in the Nazi mold, what they called "German" (*deutsche*) or "Aryan" physics. This ideology, never fully articulated so as to allow easy adaptation to new circumstances, was aimed, in their words, at freeing the physics profession from so-called Jewish-Marxist domination by the appointment of themselves and their supporters to leading positions. In so doing, they promised to free physics itself from "great dogmatic theories . . . Einstein's relativity theories, Heisenberg's matrix theory, and Schrödinger's wave mechanics," by reemphasizing their own experimental and applied research.[19] Their dogma had at its center the common Nazi and fascist notion that all human creations, even science, are expressions of an individual's race and ethnic origin. Aryan physicists produce so-called Aryan physics, that is, classical experimental physics, "the physics of those who have founded natural research," as Lenard put it. Non-Aryans, he argued, produce its antithesis, "Jewish physics," as exemplified by the theoretical physics of Einstein, Sommerfeld, Planck, Heisenberg, and the entire leadership of modern physics, whether of Jewish heritage or not.

"Aryan physics" and Lenard's and Stark's influence essentially destroyed the German physics profession. Within months of Hitler's rise to power in 1933, the Nazi state drove most Jewish and leftist professors and teachers, all of whom were civil servants, from their jobs and into exile. Any of those who decided to stay in private research (there were no private universities in Germany) were subjected to mounting indignities. Einstein, on a visiting professorship to the California Institute of Technology, immediately resigned from the Prussian Academy and never set foot in Germany again. He tried unsuccessfully to rally democratic nations against the Hitler regime. To Wilhelm Herzog, a German author, he wrote in 1933, "The great task [of German intellectuals in exile] is to impress upon those countries that still adhere to the ideals of culture and freedom, the living conviction that it is terribly dangerous to witness events as passive spectators. . . . Otherwise, within a few years, much sacrifice of human life will be unavoidable, and, even then, there will be no certainty of a favorable outcome."[20] Soon afterward Einstein, the committed pacifist, declared pacifism to be untenable in the face of Nazism.

By helping to drive modern physicists into exile, Lenard, Stark, and their cronies contributed to the rise of physics in other nations, such as the United States. Their influence waned only in the depths of World War II, as Germany found it necessary to turn to modern science for new weaponry and industrial processes; it ended only with the destruction of the Third Reich.

IDEALISTS

The German physicists Lenard and Stark had mingled their personal prejudices regarding relativity theory with the dominant strains in the political culture of their time and nation, that is, with the rising fortunes of Nazi ideology. In essence, they had exploited the reigning ideology and public anxiety over relativity theory to advance their own views and careers at the expense of relativity theory in particular and German theoretical physics in general. Unfortunately, their behavior was not unique among physicists, although it was among the most extreme. Similar instances have occurred involving leading physicists in at least three other widely diverse political settings: the Soviet Union, Communist China, and the Anglo-American cultural milieu. Such wide diversity indicates that it was not relativity theory itself—which these physicists were quite capable of understanding—that was most influential but what these physicists made of the theory for their own advantage, as they presented a distorted image of it to a demanding public in their respective political environments.[21]

Historian Loren Graham has examined the Anglo-American and Soviet settings. He has found that events in the latter occurred in large part as a response to events in the former.[22] The difference may be reduced to the old philosophical difference between the materialist and idealist conceptions of nature. Materialists hold that at the foundation of nature is matter moving in absolute time and space subject to objective physical laws, which, like nature itself, are independent of human existence. Idealists maintain the opposite: the human mind, not matter, is fundamental, since it imposes order, natural laws, and ultimate reality upon nature.

British physicists James Jeans and Arthur Eddington were among the most prominent and widely read popularizers of relativity theory in England and the United States. Their portrayals of the theory were decidedly idealistic. "Without the mind there is but formless chaos," Eddington declared. "The stuff of the world is mind-stuff." Jeans conjectured in 1934 that the ultimate understanding of nature would result in "the total disappearance of matter and mechanism, mind reigning supreme and alone."[23]

Eddington told his readers that relativity theory was, as Einstein himself had suggested, a free creation of the human mind, a theory abstractly deduced from subjective, a priori principles and applied to the real world through "thought experiments" and a small amount of elementary empirical data. Therefore Eddington was not willing to draw parallels outside of physics. Instead, he declared, the new theory applied only to a limited realm of physical nature and certainly did not touch on human ethics and religion. As a result of this position, Eddington could reassure an unsettled public that traditional ethics and religion would remain unharmed by the new physics.

In a remarkable statement to the contrary, Eddington's colleague James Jeans went so far as to inform the public that the loss of deterministic mechanism in modern physics (both relativity and quantum mechanics) *did* have an impact on human affairs: it ensured the restoration of human free will: "The classical physics seems to bolt and bar the door leading to any sort of freedom of the will; the new physics hardly does this. . . . The old physics showed us a universe which looked more like a prison than a dwelling-place. The new physics shows us a universe which looks as though it might conceivably form a suitable dwelling-place for free men, and not a mere shelter for brutes—a home in which it may at least be possible for us to mould events to our desires and live lives of endeavour and achievement."[24]

Among its unusual allusions, this statement contains strong elements of antiscientific romanticism. It has really nothing at all to do with physics and everything to do with public anxieties and cravings. It paints an inviting picture, a new view of the world as a "home" for "free" human beings who are not mere "brutes"—a likely reference to humans as the supposed descendants of apes in Darwinian evolution. It was a picture that the British and US public eagerly received, for what they found in both Eddington and Jeans was *reassurance*. After decades of heady debate over science and religion in the wake of evolution, the Anglo-American public could be relieved to know that relativity's limited scope (Eddington) or the new worldview that it opened onto freedom (Jeans) left plenty of room for traditional religion, morals, and individualistic democracy. The new idealist physics even seemed to chase materialism, mechanism, and deterministic evolution, so frighteningly contradictory to contemporary spiritual values, into stark retreat.

MATERIALISTS

Graham and other historians have suggested that the very success enjoyed by the idealist interpretation of relativity theory in Western Europe and the United States encouraged a backlash in the Soviet Union, where the dominant social ideology was not idealism and individualistic democracy but Marxist "dialectical materialism."[25] Like others who insisted on applying physics to society, those who held to Marxist-Leninist doctrine understood nature and society in terms of one worldview or philosophical system. This system consisted of traditional materialism: nature consists of matter in motion subject to objective physical laws. But a new and fundamental dynamic process is added: the "dialectic," or the constant dynamic tension between a thesis and its antithesis, out of which a new synthesis is formed. Science and other human creations are—contrary to Nazi ethnic dogma—products not of ethnic heritage but of the material, economic relations of a society.

Ironically, both Marxist and Nazi ideologies, though so opposite in outlook, led their most dogmatic followers to react to relativity theory in the same way: with suppression of the theory and persecution of its followers. For Marxists, "bourgeois, capitalistic" society naturally produced bourgeois, idealistic forms of science, while "proletarian, socialist" society naturally produced materialistic, proletarian, and progressive forms of science. If relativity was being celebrated as a triumph of idealism and individualism in the "bourgeois" West, it had to be banned in the "proletarian" Soviet Russia. During the 1920s Soviet demagogues used Lenin's widely read rejection of Mach's critical philosophy as bourgeois idealism to persecute adherents of relativity theory, whom they castigated as followers of Machian idealism.

Nevertheless, the fate of relativity theory in the Soviet Union took a more positive turn from that suffered in Nazi Germany. As adherents of relativity theory began to "disappear" under Stalin, several prominent physicists managed to rescue relativity by rendering it more compatible with dialectical materialism. One of the most prominent of these physicists was V. A. Fock. As Graham points out, he, like others elsewhere, exploited the dominant ideology to promote his own views, but those views entailed the promotion of relativity theory as an essential part of modern physics. Still, as a convinced Marxist, he and others found it necessary to distort certain aspects of the theory in order to squeeze it into the Marxist mold. In particular, they revived Minkowski's mistaken notion of absolute space-time as the seat of observed events and argued the existence of a uniquely preferred reference frame (located at infinite distance) in the theory of general relativity. Some of these aspects opened up new areas of research. But world war, nuclear weapons, and the race with the West for scientific and technological supremacy carried the day and encouraged the practical utilization of theories such as relativity that had long since been shown to yield useful results, whatever the ideological debates in learned circles.[25]

An entirely different ideological exploitation of relativity theory occurred in Communist China. As in the above examples, much research remains to be done. According to one author, Mao Zedong, the leader of the Communist Revolution, became an early aficionado of relativity theory, quantum mechanics, and "Einsteinianism" through Western influences and visitors to China during the 1920s. Tutored by Marxist Japanese physicist Sakata Shoichi, who packaged Einstein "in metaphors that were Mao's too," Mao became a committed political "Einsteinian."[26] For Chairman Mao, as for others elsewhere, this actually had nothing to do with science but everything to do with politics; only analogies, similes, and metaphors linked the two.

Knowledge about political affairs must arise, Mao believed, as knowledge about nature occurs in quantum mechanics: through the direct involvement of the observer. The changes in science wrought by Einstein's physics seemed a metaphor for the continual changes in human affairs wrought by the Marxist dialectic. To Mao this meant that there were no political absolutes—not even Marxism or Maoism. The Chinese revolution would be a perpetual revolution, a perpetual shifting of absolutes, even a perpetual smashing of power blocs like the smashing of atoms by elementary particles in a high-voltage accelerator.

The elementary-particle simile, Edward Friedman argues, contributed to Mao's decision to split the "nucleus" of the Communist Party leadership by unleashing the infamous Cultural Revolution in 1966. The origins of the Cultural Revolution remain murky, and the treatment of Einstein and relativity remains in many ways a puzzle. Both had enjoyed favor under Mao, yet the radical minions he

unleashed in the universities and academies of science were decidedly antiscientific and anti-Einsteinian. Einstein was attacked as a
reactionary capitalist, and his relativity theories were denounced as
bourgeois "relativism." Materialist "truth" must be objective,
absolute, and proletarian—views that Mao himself apparently once
held regarding relativity.

Numerous scientists (and many others) suffered profound
indignities under the revolutionary rampage of the Cultural Revolution, and the teaching of relativity theory and quantum mechanics
came to a virtual halt in Chinese universities. Not until after Mao's
death in 1976 and the arrest of the Gang of Four soon afterward
were Einstein and modern physics restored to prominence in China.

CULTURAL
RESONANCES

Study of the influence of scientific developments on cultural creations is still in its infancy, not least regarding twentieth-century physics and Einstein's theories of relativity. References to such influence abound across a variety of disciplines, but there is no consensus as to how precisely the new physics impinged upon each discipline individually and upon the broader culture as a whole.

Part of the reason for this is that, for those without a specialist's understanding of relativity theory, there was no consensus as to what the new physics actually entailed. Although excellent popularizations existed early, including at least two by Einstein, many readers lacked the confidence to consult them. Others, working in fields to which Einstein and most educated people could enjoy ready access, found the necessity of consulting popularizations rather than Einstein's original works a humiliating experience. Some despaired of ever understanding what Einstein said. The experience was expressed by

Lionel Trilling in 1973: "Physical science in our day lies beyond the intellectual grasp of most men. . . . This exclusion of most of us from the mode of thought which is habitually said to be the characteristic achievement of the modern age is bound to be experienced as a wound given to our intellectual self-esteem." To this, historian Fritz Stern adds, "I felt this exclusion the more as I came to realize the intensity of the aesthetic joy that Einstein and his colleagues found in their discoveries, as their correspondence exemplifies. We are shut out from that knowledge and that particular beauty."[1]

The exclusion of nonspecialists from technical science is nothing new: ancient and medieval scientists deliberately withheld their discoveries from the public, and seventeenth-century Newtonian science, though widely disseminated via the printing press, was and remains fully accessible only to the technically initiated. But what is new in twentieth-century reactions to Einstein is the profound sense of anxiety, alienation, and even resentment that exists across the divide of the "two cultures" and that runs through many of the cultural reactions to relativity theory, this central component of the incomprehensible new physics.

Some of these feelings found expression in cultural disciplines far beyond physics, and they were often reflective of reactions to relativity within the broader political culture. However, as discussed in the previous chapter, it is not the understanding of relativity theory, nor the lack thereof, nor feelings associated with either, that exerted the greatest impact outside of science, but the metaphorical "resonances" that Einstein's work elicited in the minds of "exalted spirits" in other disciplines, leading to new creative expressions of the human experience.

One common theme that will emerge from the following overview of the impact on several disciplines is the preoccupation that relativity theory seemed to engender in every field regarding two fundamental notions: the concern for *meaning* in all facets of human experience, and the replacement of static processes and events by dynamic ones owing to a new preoccupation with the concept of *time*.

PHILOSOPHY

If the "broader culture" is considered to include all intellectual or creative endeavors beyond science and technology, probably the earliest cultural response to relativity theory occurred among philosophers. That Einstein's theories of relativity possessed profound implications seemed obvious to philosophers, both as professionals and as citizens of their times. The connection was so close, in fact, that some philosophers claimed Einstein as one of their own. For instance, in 1949 the renowned Library of Living Philosophers published an important two-volume edition of essays on Einstein and his work, titled *Albert Einstein: Philosopher-Scientist.*

While most philosophers refrained for lack of expertise from rendering any professional response at all to Einstein and relativity theory, a few could rely upon extensive training in physics and mathematics. Philosophical responses to relativity theory (not all by professional philosophers) fell in the early years along the traditional lines of idealism and materialism. These interpretations were mainly for personal and public—not professional—consumption. As such, they reflected the personal and public needs of the times. As seen in the previous chapter, the pursuit of spiritual and political reassurances of various types led in the Anglo-American sphere to idealist interpretations by physicists such as Jeans and Eddington and in the Soviet Union to interpretations by physicists and philosophers along the lines of dialectical materialism.

These traditional interpretations soon gave way among professional philosophers to new directions that spread throughout twentieth-century philosophy. These can be traced to analyses of Einstein's 1905 relativity paper and especially to his analysis of the concept of the simultaneity of two events. One direction, noted by A. P. Ushenko, entailed the preoccupation with time, the shift (he writes) from the age-old "metaphysics of material and mental things" to the "metaphysics of events." Different spatially separated parts of an object would appear different to different observers because perceptions of different parts cannot be simultaneous to all observers.

Indeed, more recently, Peter Galison has argued that the practical problem of synchronizing clocks over great distances, such as is required in running an early twentieth-century train system on time, had inspired Einstein, the Swiss patent clerk, to consider the actual process of establishing the simultaneity of two events, such as the arrival of a train in a station as a clock registers the arrival time.[2] Beyond the "unification of time" so prized by European station masters, relativity theory indicated that space and time are no longer separate but united into a single entity, space-time. This means that a stationary object, such as a train stopped on a track, is no longer just a thing, but, because of the time dimension, it is a thing moving through space-time; that is, it is an *event*. The clock is still ticking, even if the wheels are not turning. To some this meant that static material and mental objects no longer possess any meaning; only events and processes do. Bertrand Russell explained: "Since Einstein . . . the physical world is no longer regarded as consisting of persistent pieces of matter moving in three-dimensional space, but as a four-dimensional manifold of events in space-time. The old view resulted from an attempt to make the common-sense concept of 'things' available for science; the new view means that 'things' are no longer part of the fundamental apparatus of physics."[3]

In his 1905 paper on the theory of special relativity, Einstein had started from two general postulates from which he deduced through hypothetical thought experiments new definitions, or meanings, of such fundamental concepts as simultaneity, space, time, and mass. The new meanings assigned to concepts that had always been regarded as absolute and given seemed to imply numerous statements about the material world, but they also spawned public anxieties about the meanings of other, nonphysical notions—such as values, morals, and human and social destiny—to which relativity theory did not at all apply. Some members of the public reacted negatively toward science for its perceived challenge to ethical traditions. Out of this mélange of concern for meaning and the public blurring of the distinction between science and human affairs (to the detriment of science) there arose one of the dominant philosophical movements of the twentieth century, one that particularly gained allegiances in the United States and Great Britain—the movement known as logical positivism or logical empiricism.

Moritz Schlick, holder of a teaching chair at the University of Vienna, had come to philosophy from physics and maintained close connections with leading physicists, such as Einstein. In 1917 he published probably the first philosophical interpretation of relativity theory. His teaching and subsequent research drew a group of brilliant young philosophers, mathematicians, and physicists to Vienna, who developed a particular philosophical approach to science. They called themselves the Vienna Circle.[4] Particularly concerned with the philosophical implications of relativity theory and, later, quantum mechanics, the Vienna Circle thrived until the mid-1930s, when its members scattered around the globe after the rise of Hitler and his annexation of Austria.

Suggestive connections exist between relativity theory and the origins of the Vienna Circle. These connections are discernible in two related themes of the Circle: the emphasis on the nature and clarification of scientific meaning and an admiration for science—especially physical science—that manifested itself in part in efforts to improve its methodology and to defend it philosophically from the reproaches of an anxious public.

With relativity theory as a model, the Circle's first order of business was to clean up the ambiguous language of science by carefully clarifying the meaning of all concepts. Following in the "positivistic," or empiricist, tradition, and inspired by Einstein's definitions of concepts such as simultaneity in terms of laboratory-style operations, the Vienna Circle held that the only statements that are meaningful in science—that is, the only ones that will contribute to positive knowledge—are those that can be verified by direct experimental evidence. These philosophers, especially those who later landed in the United States, accorded US physicist Percy W. Bridgman's "operationism" especial notice, in which scientific concepts are defined by the practical experimental operations with which they are used and measured. Like the "events" of the new physics, meaning comes through acts of observation. "The final giving of meaning always takes place therefore through *deeds*," wrote Schlick. "It is these deeds or acts which constitute philosophical activity."[5]

This emphasis on deeds and empirical meaning was tailor-made for a full-scale assault on "metaphysics," that is, upon all statements regarding the meaning of concepts of concern to the public but beyond the realm of physics, such as morals, ethics, religion, and human affairs. The spokes of the Vienna Circle argued that they did not wish to deny the importance of these wider issues, only to deny their association in any way with science, so as in large part to protect science from public disdain. While others mingled science with religion, politics, and ethics, the Vienna Circle declared all extra-scientific "metaphysics" to be nonempirical, hence scientifically "meaningless." Rudolf Carnap, a leading member of the Circle, proclaimed, "Logical analysis, then, pronounces the verdict of meaningless on any alleged knowledge that pretends to reach above or behind experience. . . . Further, the same judgment must be passed on all *philosophy of norms*, or *philosophy of values*, or any ethics or esthetics as a normative discipline."[6]

In essence, the Vienna Circle had come to the defense of science and relativity theory by retreating from the very activity that the public had for centuries demanded even more of philosophy than of physics: guidance on such metaphysical issues as systems of belief and virtuous behavior. By the crucial support that it rendered for relativity theory and physics as a discipline, the Circle had in a sense turned on itself, and on philosophy in general, in the public eye. For that reason, after scattering abroad, the Vienna Circle and its antimetaphysics program never regained its original influence in central Europe, nor did it ever spread to Communist countries, where Lenin had long since condemned Machian positivism and "empiriocriticism" as bourgeois idealism. The main setting of the school has been in the United States and Great Britain, where, after World War II, it was gradually absorbed into mainstream analytic philosophy.

A second important component of logical positivism was its emphasis on symbolic logic. Heavily influenced by Ludwig Wittgenstein and Bertrand Russell, and enamored of Einstein's theory of relativity, the Vienna Circle attempted to re-create the logical structure of a model theory, such as special relativity. This was at first intended for descriptive purposes only, but it later became "prescriptive" and "normative" for all emerging scientific theories.

The two components of relativity theory—hypotheses and empirical statements—seemed characteristic of any theory. They had to be linked in a precise, logical fashion that could be reduced to symbolic statements and rules. Again the Circle rose to the defense of physical science by demonstrating the coherence and power inherent in the logical formulation of theories, but here again the logical program of the Circle rolled into a thicket of thorny problems. Not only did it prove impossible to separate wholly empirical and theoretical statements from each other, but actual scientific research simply did not and would not follow the "normative" constraints of the positivists. Adherents of logical positivism again found themselves isolated from their audience.

The history of science, rising to prominence as a profession amid public fear of science and technology following World War II, became instrumental in pointing out the disparity between science as it is actually practiced and the logical ideal preached by physicists and philosophers. A leading critic of the logical positivist interpretation of the history of science was Thomas S. Kuhn, a philosopher and former physicist. His famous book *The Structure of Scientific Revolutions*, first published in 1962, stressed the significance of such nonlogical elements in the development of science as psychology, group dynamics, dogmatism, and individual perceptions.[7] In a new and anxious age in which physics seemed to have succeeded more than any other intellectual endeavor in determining human destiny, some felt that the time had come to bring the "free-floating" ideal, logical interpretation of science back to earth as a historical and cultural phenomenon.

Logical positivists, and some close to that school such as Karl Popper, responded to Kuhn and the historians by arguing that historians must distinguish the "context of discovery" from the "context of justification": Einstein's actual path to the formulation of relativity theory may not have followed logical precepts, but the acceptance of his theory surely entailed logical elements. Even if that failed, the steps in the formulation and acceptance of the theory could be retrospectively recast in logical terms, as an abstract "rational reconstruction."[8] Despite such theoretical debates, which continue today, in going about their work most historians of science follow neither Kuhn's ideas nor any other logical or alogical interpretation of the history of science.

The aim of most historians of science is no longer to defend or to recast past achievements such as Einstein's in terms acceptable to the present but to comprehend the past in its own right: what actually happened and why it occurred then in that place through those individuals. Einstein's relativity theories are seen by historians, and now also by many philosophers, in ways similar to other scientific developments: as historical and cultural phenomena. As such, they are subject to scrutiny through methods and approaches closer to those employed by social and cultural historians and philosophers than to those provided by logical reconstruction or by any theory of history or of science.

VISUAL ART

If art reflects life, the life of the early post–World War I years was one of unusual upheaval as well as creativity. Art, like science and philosophy, struck out in new directions, but it also reflected the anxiety of its times, and meaning and process were again central issues. One artistic school, known as dada, though beginning during the war, reflected to an extreme the despair and disillusion of those who, in the throes of the war and its aftermath, believed European culture had lost any meaning that it once held. Dadaists reacted to the loss of meaning in nihilistic, antiart artistic outbursts. While some nonartists opposed relativity theory by calling it "scientific dadaism," dadaists depicted scientists as speaking meaningless gibberish. One dadaist jabbered in an outburst of stream-of-consciousness, "Logic strained by the senses is an organic illness. The philosophers like to add to this element: The power to observe. But, as a matter of fact, this magnificent quality of the mind is the proof of its impotence. We observe, we look from one or more points of view,

we choose them from among the millions that exist. . . . Science disgusts me as soon as it becomes speculation-cum-system, loses its
useful character—which is so useless—but at least individual. I
loathe fat objectivity and harmony, this science that finds everything
in its place. Carry on kids, humanity."[9]

On a less garbled note, the parallels between relativity theory
and visual art during and after World War I are striking. They are all
the more so in that, during the first decades, both sides of the parallel vigorously denied any direct influences, and, in most cases,
none seem to exist. One of the most striking parallels occurred in
the school of painting and sculpture known as cubism. One art historian, who attempted to draw the connections more closely in an
essay titled "Cubism and the Theory of Relativity," prompted a terse
note from Einstein: "This new artistic 'language' has nothing in
common with the Theory of Relativity."[10]

Frustratingly, the parallels (if not the connections) do exist.[11]
One critic, describing the analogy between Einstein's work and a
school of recent sculpture, found it necessary to interject: "The
analogy stops short of physical reality and rests upon visual correspondences. Precisely how these correspondences came about is difficult to say."[12] Perhaps all that can be said at this point is that trends
in creative ideas apparently do exist at certain times, infusing and
transcending individual creations and resonating throughout an
entire historical epoch.

Like relativity theory, the roots of cubism stretch long before the
early work of Einstein or of the founders of cubist painting, Pablo
Picasso and Juan Gris. Yet both emerged through the near-simultaneous discovery in both fields of new conceptions of space, time,
and observation. Like Newtonian physics, Renaissance painting
depicted a single scene in a static, absolute space—a snapshot in
which everything is simultaneous and viewed from one perspective.
Cubism, to the contrary, apparently deriving from Cézanne's
"geometrization" of nature, introduced a shifting point of view in
which a single figure is seen from different sides, as if the eye moved
through a sequence of relative perspectives, none of which can be
considered privileged. These multiple perspectives were intended to

enable the artist to grasp the essence of the object by breaking it into geometrical planes of visual and emotional insight.

The introduction of relative perspectives can be interpreted as the introduction of processes or sequences of events (as in philosophy), as the loss of simultaneity, as the introduction of a fourth dimension—time—to the three-dimensional space of traditional painting. And, as relativity theory transcended Newtonian physics, the new "geometrical" painting broke the bonds of the Renaissance canvas, opening up new avenues of experimentation with space and time, as in Marcel Duchamp's famous *Nude Descending a Staircase* (1912). Poet Guillaume Apollinaire wrote in a 1917 essay "On Painting,"

> The new artists have been violently attacked for their preoccupation with geometry. Yet geometrical figures are the essence of drawing. Geometry, the science of space, its dimensions and relations, has always determined the norms and rules of painting. Until now, the three dimensions of Euclid's geometry were sufficient to the restiveness felt by great artists yearning for the infinite. . . . Today, scientists no longer limit themselves to the three dimensions of Euclid. The painters have been led, quite naturally, one might say by intuition, to preoccupy themselves with the new possibilities of spatial measurements which, in the language of the modern studios, are designed by the term: the fourth dimension.[13]

By its very nature sculpture, which deals with multiple dimensions, multiple perspectives, space, and mass, also displayed striking parallels with the new physics. Matisse and others attempted to realize in sculpture the temporal and geometrical ideals of cubist painting. One of the most successful pieces was Jacques Lipchitz's *Man with Mandolin* of 1917. "As with Cubist painting," writes one commentator, "there is now no single point of view, no continuity or simultaneity of image contour."[14]

But something new had also occurred in Lipchitz's sculpture that blossomed in many different directions in the decades ahead: not only was the traditional manipulation of mass in sculpture handled in a new way, but there was a so-called piercing of the mass, an introduction of empty space into the sculptor's domain. As in general relativity theory, where space and mass are conjoined, space now joined mass as an object of the sculptor's craft. As the pioneers of this "new reality" put it in 1920, "Mass and space are two concrete and measurable things. We consider and use space as a new and absolutely sculptural element, a new material substance which really enters into construction." Moreover, they declared, "kinetic and dynamic elements must be used to express the true nature of time."[15]

Excellent expressions of the "new reality" may be seen in Naum Gabo's *Kinetic Sculpture: Standing Wave* (1920) and his *Linear Construction* (1950), in which space seems captured by the fields of force created by plastic and nylon threads. The field concept of relativity physics is reflected even more in the works of the post–World War II school known as New Tendency. Here space itself is replaced by fields of temporally changing energy achieved, as field theorists would have it (and as contemporary urban existence implies), by arrays of individual spatially distributed points of no consequence in themselves but only in their entirety.[16]

LITERATURE

Einstein's influence on literature has received even less attention than that accorded his influence on art and philosophy. But that influence was in some cases more direct and noticeable than it was in art. Again, meaning and process are central. Indeed, some have attempted to trace the preoccupation with meaninglessness in the works of the absurdist school of fiction—notably Albert Camus and Jean-Paul Sartre—directly to Einstein and Heisenberg. As in nearly every other cultural field, some people extended relativity beyond its breaking point by arguing that—since all points of view are equally valid—ultimate meaning is relative, and absolute knowledge dissolves into subjectivity. Not only is such an interpretation of relativity theory tantamount to the false claim that Einstein asserted "everything is relative," but, like dadaism, the origins of schools such as absurdism lie deeper in the human experience of a sometimes brutal and bewildering century than in any scientific theory.

Gerald Holton has pointed out other examples of severe misunderstandings of Einstein among authors and literary critics. But certain novelists—whether they understood Einstein or not—produced works that, consciously or unconsciously, seemed to resonate in their own right with the new physics. One example cited by Holton is William Faulkner's *The Sound and the Fury*. As in a cubist painting, time and process seem to merge into the events as they unfold in Faulkner's novel.[17]

Ironically, an even more direct influence may be seen at precisely the spot where the divide between the "two cultures" is often believed to be the widest: between physics and poetry. Because of its rational, mathematical structure, physics is usually regarded as wholly incompatible with the intensive immersion of poetry in the emotional array of human feeling and experience. One physicist, Heisenberg, recalled an early encounter with advanced geometry: "I felt this to be very dry stuff; triangles and rectangles do not kindle one's imagination as much as do flowers and poems."[18] (His imagination was soon drawn to geometry, nonetheless.)

The tension between poetry and physics was always present, however, despite any positive connections between the two. For this very reason, such leading English-language poets as Pound, Frost, and e. e. cummings displayed both skepticism and admiration regarding Einstein's science. According to one commentator the skepticism derived less from reaction to Einstein than to the antihumanistic—hence antipoetic—uses that the public was making of his science.[19] Such skepticism may be seen in cummings's 1931 satire of the new science (note his pointed use of upper- and lowercase letters):

Space being(don't forget to remember)Curved
(and that reminds me who said o yes Frost
Something there is which isn't fond of walls)

an electromagnetic(now I've lost
the)Einstein expanded Newton's law preserved
conTinuum(but we read that beFore)

of Course life being just a Reflex you
know since Everything is Relative or

to sum it All Up god being Dead(not to

mention inTerred)
                    LONG LIVE that Upwardlooking
Serene Illustrious and Beatific
Lord of Creation,MAN: . . .[20]

Other poets saw in Einstein something quite different: a man who, like themselves, sought and achieved new insights into nature. The new discoveries that Einstein had brought to physics, they felt, should be incorporated into poetry. Among the leaders of this school, known as "objectivism," were the noted poets Archibald MacLeish, William Carlos Williams, and Louis Zukofsky. Each made serious and successful efforts to understand relativity theory. Zukofsky apparently became converted to this school after translating a biography of Einstein by Einstein's son-in-law into English.

Aside from tributes to Einstein and his work—such as Williams's celebration of Einstein's visit to the United States in 1921 titled "St. Francis Einstein of the Daffodils"—what did these poets make of relativity theory? The most obvious answer may be seen in the most obvious feature of any poem: its measure. "The only reality that we can know is MEASURE," wrote Williams, and, he declared, poets could not accept relativity theory "without incorporating its essential fact—the relativity of measurements—into our own category of activity: the poem." Like Williams, numerous poets equated the relativity of measurements with the relativity of poetic measures. All sorts of experimentations with verse, structure, and meter, already underway, continued to expand, rendering for many the form and

---

content indistinguishable. "I say we are *through* with the iambic pentameter as presently conceived, at least for dramatic verse," Williams proclaimed.[21] Zukofsky arduously experimented in the many versions of his poem "A" to render the poem a type of mathematical equation. Williams, in perhaps his best-known poem, "The Red Wheelbarrow," stripped not only the meter but the words to their bare observed essentials, achieving a type of field effect in which words and images blend in numerous marvelous and diverse ways.

Archibald MacLeish's long poem "Einstein," published in 1929, has been called "the finest poetic tribute to the scientist." It is perhaps also the most exhaustive exploration of the significance of Einstein's physics in both the form and content of any poem. MacLeish's epic is heavily introspective and personalized. "The chief idea to emerge from the poem," writes Hyatt Howe Waggoner, "is that the new Einsteinian science has at once emphasized the centrality and increased the loneliness of the knowing mind."[22]

As always, in poetry and in culture, the tension between nature and the knowing mind remains. Einstein, isolated yet knowing, is confronted with himself and the discovery that nature harbors "something inviolate." It is a something that returns in the last lines of MacLeish's epic to haunt Einstein—and us—at the end of life:

> *Einstein enters*
> Like a foam
> His flesh is withered and his shriveling
> And ashy bones are scattered on the dark.
> But still the dark denies him. Still withstands
> The dust his penetration and flings back
> Himself to answer him.
> Which seems to keep
> Something inviolate. A living something.[23]

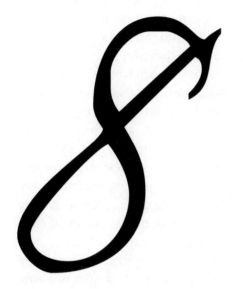

SCIENCE
AND
CONSCIENCE

What of the living man, the frizzy-haired nonconformist, the subject of innumerable press photographs who seemed ready to offer an opinion on everything from elementary education to nuclear weapons? In what ways did he personally influence our world? The answer derives from Einstein's intense personal devotion to democracy, economic justice, human dignity, and intellectual freedom. Such devotion was coupled with his equally strong sense of moral responsibility as a leading public scientific figure to speak out in defense of these ideals in an era of totalitarian regimes, persecution of dissent, and weapons of mass destruction.

Here again much remains to be learned about Einstein's views and their broader influence. But insight can be gained from the confluence of the two in an early defining event of the twentieth century: World War I. Einstein did not begin to engage publicly in social issues until after the outbreak of war, and his views achieved universal notice only amid the turmoil following that war. Like Bacon, Descartes, Spinoza, and others before him, Einstein regarded intellectuals, especially scientists, as a cultural and moral elite. As researchers into the secrets of nature, they represented all of humankind. Because of this, they bore a special responsibility to preserve the best of the human cultural tradition. Nationalism, militarism, and exploitation not only undermined human dignity, but they were also the root causes of inhuman war and violence. Cultural development and academic research were two of the truly transnational endeavors that could work against these negative factors. For Einstein, the increased exploitation of science for military power and other self-serving ends as the century progressed greatly increased the burden upon scientists to speak out as voices of conscience and moral stability in an era of political turmoil.[1]

Despite his many pronouncements that his science had nothing to do with human affairs, for Einstein the scientist as human being had everything to do with them. Many of Einstein's German colleagues did not agree. Soon after Hitler's rise to power in Germany, Max Planck and Max von Laue complained to Einstein, then in the United States, that his strong pronouncements against Germany not only violated the tradition of the "apolitical professor"—the silent withdrawal of intellectuals from the "dirty world" of politics—but they were also being used against those who remained in Germany. Einstein defended his actions to Max von Laue, a courageous man in his own right, in unmistakable terms: "I do not share your view that the scientist should observe silence in political matters, i.e., human affairs in the broader sense. The situation in Germany shows whither such restraint will lead: to the surrender of leadership, without any resistance, to those who are blind or irresponsible. Does not such restraint signify a lack of responsibility? Where would we be, had men like Giordano Bruno, Spinoza, Voltaire and Humboldt thought and behaved in such a fashion? I do not regret one word of what I have said and am of the belief that my actions have served mankind."[2]

With Hitler in power, Einstein renounced his German citizenship for the second time (but he maintained his Swiss citizenship, later becoming also a US citizen). He remained a resident of the

United States for the rest of his life, settling at the Institute for Advanced Study in Princeton, New Jersey.

The public generally supported Einstein in the role he assumed as a public figure. After all, broad reaches of post–World War I society attributed to Einstein, a member of a newly emerged elite of theoreticians, the discovery of new universal truths that went far beyond science. How one reacted to these truths often determined how one reacted to his social message. As political figures, as well as demagogues and fanatics of various persuasions, competed for public allegiance, Einstein became a type of lonely prophet: a refreshing alternative for some, an annoying gadfly to others.

What was Einstein's message to the public? In essence it was a defense of human dignity and freedom as the highest goods. During and after the world war he became an outspoken supporter of pacifism, internationalism, constitutional democracy, and the egalitarian distribution of wealth, leaning toward a socialist economy. These were not idealistic or simplistic stances, as is often asserted, but positions informed by a realistic assessment of human nature and its propensity for greed, hatred, and lust for power. Because of this, he was never dogmatic in his adherence to social positions but rather always willing to adapt his views on such fundamental issues as pacifism and national identity to changing circumstances when he felt that his ideals would be better served—often to the consternation of his supporters. Such willingness to adapt was most evident in two of his most well-known causes: his pacifist opposition to war and his support for the founding of Israel.

EINSTEIN AND ISRAEL

As political philosopher Isaiah Berlin has pointed out, Einstein always detested the barriers of nation, state, and ethnicity that separate people from one another. He saw himself as a universal citizen of the world.[3] But nationalism, ethnic identity, and separation were on the rise in the era of World War I. As German chauvinistic nationalism and anti-Semitism increased after the war, especially in Berlin and against him personally, Einstein drew ever closer to the growing Jewish national movement known as Zionism. But even earlier, upon moving to Berlin, he was suddenly reminded of Jewish cultural identity and of his own heritage. In 1929 he wrote to a German minister of state, "The Jews are a community bound together by ties of blood and tradition, and not of religion only: the attitude of the rest of the world toward them is sufficient proof of this. When I came to Germany fifteen years ago, I discovered for the first time that I was a Jew. I owe this discovery more to Gentiles than to Jews."[4]

Einstein was practically alone among European Jewish intellectuals in his early realization of the precarious position of Jews in Europe. He concluded that the assimilation of Jews in Western Europe had failed. "However much the Jews adapted themselves, in language, manners, and to a large extent even in the forms of religion, to the European peoples among whom they lived, the feeling of strangeness between them and their hosts never vanished. . . . Nationalities want to pursue their own goals, not to blend."[5]

As a result of such realizations, Einstein supported the growing movement for a Jewish national homeland in Palestine led by Chaim Weizmann, a Russian chemist who had become a British citizen. Einstein's aim in supporting the Zionist movement was to help create a Jewish national and cultural identity rather than a new Jewish nationalism. As Berlin put it, "He hated nationalism all his life. But he recognised the acute need of the Jews for some form of national existence."[6] By the same token, as a German professor subjected to anti-Semitism, he also immediately recognized the need for a Jewish academic institution in which Jewish scholars could work undisturbed. As early as 1918 he lent his support to the newly founded Hebrew University of Jerusalem as "a spiritual center for the Jews of the whole world."[7] In April 1921 Einstein made his first trip to the United States in the company of Weizmann for the purpose of raising funds for the new university. The mission proved a huge success, and the university was opened in 1925. In 1927 Einstein willed his papers to the Hebrew University, where they now reside.

With Hitler's rise to power in Germany and the failure of the Western democracies to oppose his internal policies and preparations for war, Einstein became even more convinced of the need for a Jewish homeland. "Palestine," he wrote in 1934, "will be a center of culture for all Jews, a refuge for the most grievously oppressed, a field of action for the best among us, a unifying ideal, and a means of attaining inward health for the Jews of the whole world."[8] But Einstein also recognized early on the inherent practical problem already evident at that time in Palestine—the relationship between the Jews and the Arabs of Palestine. Tensions were rising, and in response to a concerned letter from a Palestinian Arab in 1930, Einstein wrote, "This state of affairs is unworthy of both nations and can only be altered by our finding a *via media* on which both sides agree."[9] But tensions persisted, and with the founding of Israel after World War II, Einstein was critical both of the British Mandate, which did not render a partition of Palestine into Jewish and Arab sectors, and of Zionist nationalists who sought to exclude the Arabs from Palestine through violence if necessary. In 1949 he declared: "One of the [ethical ideals of our people] is peace, based on understanding and self-restraint, and not on violence. If we are imbued with this ideal, our joy becomes mingled with sadness, because our relations with the Arabs are far from this ideal at the present time."[10]

Chaim Weizmann, the first president of Israel, died on November 9, 1952. A Tel Aviv newspaper (*Maariv*) immediately proposed Albert Einstein, the most famous living Jew, as Weizmann's successor. On November 16, David Ben-Gurion, the premier, instructed Abba Eban, then ambassador to the United States, to offer the position to Einstein. Einstein, however, learned of the offer in advance and was already prepared with an answer. He had decided to refuse, he wrote to Eban, because of his age and health, his lack of experience in administrative matters, and especially because of "the difficult situation that could arise if the government or the parliament made decisions which might create a conflict with my conscience; for the fact that one has no actual influence on the course of events does not relieve one of moral responsibility." Nevertheless, the man who had once declared himself a complete loner, wrote to Eban in 1952, "I am the more distressed over these circumstances because my relationship to the Jewish people has become my strongest human bond, ever since I became fully aware of our precarious situation among the nations of the world."[11] Einstein died three years later.

EINSTEIN AND NUCLEAR WEAPONS

Einstein had very little to do with the creation of nuclear weapons, but he had a great deal to say about them. Einstein was particularly distressed by his association in the public's mind with the dropping of the atomic bombs on Japan in 1945, resulting in the deaths of hundreds of thousands of civilians.

Einstein insisted on several occasions that his only contributions to the development of these terrible weapons of mass destruction— if they can be considered contributions—were the fundamental scientific insight that mass is equivalent to energy by the formula $E = mc^2$ and the letter he signed to President Franklin D. Roosevelt on August 2, 1939. The letter was drafted by Leo Szilard, a Hungarian refugee physicist, and signed by Einstein while on vacation on Long Island. It stated: "Certain aspects of the situation [regarding nuclear energy] seem to call for watchfulness and, if necessary, quick action on the part of the Administration." Nuclear fission, the letter explained, had been discovered in Berlin less than a year earlier, enabling (by Einstein's own formula) the release of "vast amounts of power." German scientists were apparently already at work on practical applications. Among the applications, the letter continued, "this new phenomenon would also lead to the construction of bombs, and it is conceivable—though much less certain—that extremely powerful bombs of a new type may thus be constructed." Einstein's letter concluded with the recommendation that Roosevelt "may think it desirable to have some permanent contact maintained between the Administration and the group of physicists working on chain reactions in America," in order to monitor developments.[12]

The letter to Roosevelt remains controversial in many respects. Some claim that actually it had only little effect on initiating government interest in research that ultimately led to the atomic bomb; others argue in hindsight that had Einstein, the pacifist, not signed this letter, there might not have been an atomic bomb, and that this letter compromised the moral position he took against nuclear weapons after the war. Still others emphasize that the letter was not actually written by Einstein but by Szilard and that it does not actually call for a government research effort on nuclear weapons but only for "watchfulness" and preparation for "quick action." Nevertheless, Einstein had signed the letter, and he had done so, he later wrote, out of fear at the time that Hitler and the Nazis might have gotten hold of such a weapon ahead of the Allies. "If I had known that these fears were groundless," he told Max von Laue in 1955, "I would not have taken part in opening that Pandora's box."[13]

What had happened to Einstein's pacifism in 1939, and on what basis could he then oppose nuclear weapons after 1945? The answers again reveal his pragmatic approach to the realization of his humanitarian ideals. Responding to a Japanese correspondent in 1953, he wrote, "I am a *dedicated* but not an *absolute* pacifist; this means that I am opposed to the use of force under any circumstances, except when confronted by an enemy who pursues the destruction of life as *an end in itself*."[14] Einstein realized, again ahead of many others, that Hitler and the Nazis fit this definition. But, sadly, even Einstein's voice of moral authority and his flexible pragmatism in its use did not prove sufficient to influence the course of events in both arenas—war and nuclear weapons—that came to dominate much of the twentieth century.

Einstein's commitment to pacifism, openly expressed during the depths of World War I, was well known. As late as 1931 he called himself a "militant pacifist." "I am willing to fight for peace. Nothing will end war unless the peoples themselves refuse to go to war," he declared in Chicago.[15] But upon Hitler's rise to power in Germany two years later, Einstein shocked his followers when he declared to Belgian pacifists that, in view of Germany's warlike intentions, "Were I a Belgian, I should not, in the present circumstances, refuse military service; rather, I should enter such service cheerfully in the belief that I would thereby be helping to save European civilization." For the rest of the decade he tried desperately to rally the rest of Europe and the world to military defense and pressure against the German menace. "If we do not act with energy," he declared, "we shall be doomed to a way of life indistinguishable from that under ancient Asiatic despotisms. Indeed, it will be even worse. . . . The essence of despotism lies not only in the fact that one man with virtually unlimited authority holds power, but that society itself becomes an instrument for enslaving the individual."[16]

Einstein's efforts were in vain, and his worst fears sadly came to pass. Hitler unleashed the German army into Poland on September 1, 1939, initiating an eventual world war of far greater brutality and barbarism than anyone could have possibly imagined. The Holocaust and the development of nuclear weaponry associated with this war have became two of the defining events of twentieth-century history that have shaped our world ever since. Einstein and others like him, shaken by the magnitude of both, had been powerless to stop either of them. To the end of his life a decade later he never forgave the Germans for what they had done.

After signing the letter to Roosevelt alerting him to the danger of nuclear developments, Einstein agreed to work as an advisor to the United States Navy on underwater explosions, but he was deliberately excluded from any association with the Manhattan Project that built the atomic bomb. The army refused him a security clearance, and the administrators of the project distrusted him for his independence and for his prior pacifist leanings.[17] Nor did government officials want any further advice from him, as the project marched to its inexorable conclusion. Einstein learned quietly of the prospect of a bomb in March 1945 and wrote another letter to Roosevelt, this one introducing Szilard, who would argue for a demonstration of the power of the bomb to Japanese authorities before it was used on Japanese cities without warning. The letter was found in Roosevelt's office after his death in April and was delayed by those advising his successor, Harry Truman. By the time Szilard was able to present his arguments, the government's advisors had already decided against any demonstration of the bomb or any prior warning.[18]

Einstein regarded the rise of Nazi Germany and the outbreak of world war as a failure of the international community to take forceful action against a rogue state. The US monopoly on nuclear weapons at the end of the war and the deepening cold war, as the United States and the Soviet Union confronted each other for world dominance, seemed to him to pose similar threats. In 1949, the Soviet Union became the world's second nuclear power, raising the prospect that nuclear confrontation would soon result in a nuclear war of immense world devastation. "The chief obstacle to peace," Einstein wrote in 1951, "is the desire for greater and greater power on the part of the two countries which have proven most successful in the military sphere: Russia and the United States."[19] Einstein and others became convinced of the need for an international organization to control nuclear weapons and to prevent future wars—a type of "world government" that would have the power to enforce peace with military action if necessary. He called upon the United Nations to move immediately toward establishing such a powerful world government before it was too late: "We know that the developments of science and technology have determined that the peoples of the world are no longer able to live under competing national sovereignties with war as the ultimate arbitrator."[20]

In hindsight, Einstein's proposal of a world government seems hopelessly naive. Experience has taught us that no nation, least of all the United States and the Soviet Union, would voluntarily give up its sovereignty to a higher authority with the power to force them into compliance. Even the loose form of international cooperation underlying the formation of the United Nations has not survived the national agendas of powerful nations demanding to have their own way in nuclear and other matters, as is still evident today.

Yet, as Einstein had also envisioned, the alternative to world government at that time was the very strong likelihood of nuclear war, which in fact nearly came to pass as the cold war reached frightening depths. In an attempt to avoid the impending disaster, Einstein served as chairman of the Emergency Committee of Atomic Scientists, which lobbied for civilian control of nuclear energy in the United States and through the United Nations. In addition, Einstein spoke and wrote widely on nuclear control and the need for a world government; he signed numerous appeals and petitions supporting these causes; he joined with other international figures, such as Bertrand Russell and

Niels Bohr, in opposition to the nuclear arms race; and he spoke directly to world leaders on the need for universal disarmament.

But such efforts could not compete with the lust for nuclear power on both sides of the cold war. The surprise explosion of the Soviet atomic bomb in August 1949 shocked US leaders and lent new impetus to the demand of some scientists and administrators for US nuclear superiority at any cost. Late in 1949, a small group of scientists led by Edward Teller began pushing for the development of an even more powerful nuclear device, the "super," or hydrogen, bomb, a fusion weapon of unlimited destructive power.[21] After considering the proposal, the government's top nuclear policy advisors, led by J. Robert Oppenheimer, the father of the less powerful atomic bomb, strongly opposed development of the hydrogen bomb. It was, they argued, not only unnecessary for national security but also inherently objectionable on moral grounds—and it might not even be technically feasible. Oppenheimer wrote for the committee majority, "The extreme dangers to mankind inherent in the proposal [of a program for the super] wholly outweigh any military advantage that could come from this development. . . . Its use would involve a decision to slaughter a vast number of civilians. . . . If super bombs will work at all, there is no inherent limit in the destructive power that may be attained with them. Therefore, a super bomb might become a weapon of genocide." The minority of two declared, "It is necessarily an evil thing considered in any light."[22]

Despite the opposition of most scientists and the government's leading nuclear advisors, President Truman ordered the development of the hydrogen bomb in January 1950. Einstein was horrified at the decision to proceed, as was most of the scientific community. To him the decision meant the triumph of a "mechanistic, technical-military, psychological attitude" over the broadly construed humanistic vision of science as a contributor to human culture. Making a rare appearance in the new medium of television, Einstein was a guest via film on a program hosted in 1950 by Eleanor Roosevelt. He warned, "The idea of achieving security through national armament is, in the present state of military technology, a disastrous illusion. In the United States, this illusion has been strengthened by the fact that this country was the first to succeed in producing an atomic bomb. . . . The belief seem[s] to prevail that in the end it were possible to achieve decisive military superiority."[23]

On November 1, 1952, the United States exploded the world's first hydrogen bomb. For the moment the country had achieved nuclear superiority. But with the launch of the first artificial satellite, Sputnik, in 1957, the Soviets once again shocked the Americans, who began a new effort to attain superiority in space at all costs. The primary concern was that the Soviets were ahead in the construction of intercontinental ballistic missiles, on which hydrogen bombs could be placed. The prospect of total nuclear destruction raining down on any part of the planet within minutes of a missile launch became a horror that, despite the end of the cold war at the end of the 1980s, the world continues to live with today.

In order to support the drive for nuclear supremacy after the end of World War II, the authorities had placed US science and society on a permanent war footing. This encouraged, Einstein complained, the suppression of public dissent and civil liberties, especially against cultural and intellectual figures. Responding to the introduction of a loyalty oath for professors at the University of California, Einstein explained, "The disease stems from the fact that one tries to base the security of the country on national armament rather than on international organization. The result is an increasing militarization of the country, which inevitably leads to the formation of a police state and, in fact, has already done so to a large extent."[24]

Militarization, paranoia, and cold-war anxiety reached their climax in the rise of Senator Joseph McCarthy, a cold warrior who exploited the fear of Communism for political gain. The "police state" became quite real for the victims of the McCarthy era, many of whom were driven from their jobs and their careers.[25] Einstein himself naturally became a prime target. With the recent release of the FBI's voluminous surveillance file on Einstein to public scrutiny, it is clear that J. Edgar Hoover, then director of the FBI, had been particularly eager to prove that Einstein was a Communist and a Soviet spy since the beginning of fission research in 1939.[26] According to historian Fred Jerome, Hoover did not even hesitate to utilize

German Nazi newspaper sources alleging his Communist connections, hardly the most objective sources of information. After the war, Hoover expended a great deal of effort to substantiate the claim of an unnamed former Gestapo agent that Einstein's office in Berlin had been used before Hitler's rise to power as a cable-drop for Soviet spies. The allegation of the former Nazi has not been substantiated, nor does it seem likely that it will be.

Fortunately, however, Hoover's Gestapo source was unwilling to testify in an action against Einstein, and Einstein himself was never subpoenaed to testify before McCarthy's congressional committee investigating "un-American activities." It was clear what Einstein's response would be to such a subpoena. In a letter to a schoolteacher called to testify, which was published in the *New York Times*, Einstein declared,

> The problem with which the intellectuals of this country are confronted is very serious. Reactionary politicians have managed to instill suspicion of all intellectual efforts into the public by dangling before their eyes a danger from without. . . . What ought the minority of intellectuals do against this evil? Frankly, I can only see the revolutionary way of non-cooperation in the sense of Gandhi's. Every intellectual who is called before one of the committees ought to refuse to testify, i.e., he must be prepared for jail and economic ruin, in short, for the sacrifice of his personal welfare in the interest of the cultural welfare of his country.[27]

McCarthy could not risk the spectacle of persecuting the great Albert Einstein, nor could the country risk the possibility that Einstein, who had twice renounced his German citizenship, would now renounce his US citizenship and leave the country.

But lesser-known people could not always withstand the power of Congress, and many of those who did appear faced economic ruin nonetheless. The McCarthy era gradually faded as higher authorities finally found the courage to oppose the man, but the toleration of dissent and criticism in the midst of national crisis remains to this day a tenuous privilege.

EINSTEIN TODAY

Einstein died in Princeton on April 18, 1955, of a ruptured aortic aneurism. During the almost fifty years since his death, public fascination with the man and his influence upon our world have hardly waned. His work on general relativity theory remains the foundation of research today on the structure of the universe and the behavior of black holes, neutron stars, and other exotic features of the cosmos. Even Einstein's mistaken introduction of the cosmological constant into his equations has recently proved useful in accounting for the puzzling accelerated expansion of the universe. And general relativity has made possible the astonishing precision of the Global Positioning System, widely available today to civilian and military users.

Einstein's work on special relativity, quantum theory, and unified field theories also continues to inspire research and practical applications. Particle accelerators must take into account the effects of special relativity as particles accelerate to speeds close to the speed of light; the invention of lasers, transistors, and semiconductor devices would not have been possible without quantum theory; and a future unified field theory will require a unification of the discontinuous quantum nature of matter with the continuous nature of general relativity. All of these contributions have shaped the world in which we now live and contributed to the culture and comprehension of ourselves as human beings living in a vast and wonderful, yet complicated and subtle, natural world.

Einstein also lived during a century that brought us the evils of world war, nuclear weapons, totalitarian dictatorship, and mechanized state genocide. Yet it has also brought us the examples of people like Albert Einstein who had the courage to hold to the ideals of human dignity and freedom and to stand up for their defense when necessary, even at great personal cost.

The world wars seem over for now, the cold war and the frightening prospect of imminent nuclear conflagration have ended, and world-threatening dictatorships are on the wane. Yet, in some ways, the world has changed little since Einstein's day, and in other ways it has even gotten worse. The former cold-war powers still maintain vast arsenals of nuclear weapons, other nations are finding ways to arm themselves with such weapons and the missiles to deliver them, and new biological and chemical weapons of mass destruction have been developed and even deployed at times. Genocidal persecution and "ethnic cleansing" are still practiced; nations still reserve the right to impose their will on others; and terrorism has become a new, inhuman horror of immense proportions.

Nevertheless, people of goodwill still continue to work for human dignity and freedom, as the words of Albert Einstein and Bertrand Russell issued on April 11, 1955, just one week before Einstein's death, still resound across the decades: "There lies before us, if we choose, continual progress in happiness, knowledge and wisdom. Shall we, instead, choose death, because we cannot forget our quarrels? We appeal, as human beings, to human beings: Remember your humanity and forget the rest."[28]

NOTES

R eferences are suggestions for further reading; they are not intended to be exhaustive of the subject.

ABBREVIATIONS

AE, *Ideas*   Albert Einstein, *Ideas and Opinions*, ed. Carl Seelig, trans. Sonja Bargmann (New York: Modern Library, 1994).

AE, *Notes*   Albert Einstein, *Autobiographical Notes: A Centennial Edition*, ed. Paul Arthur Schilpp (LaSalle, IL: Open Court, 1991).

AE, *Papers*   Albert Einstein, *The Collected Papers of Albert Einstein*, multiple volumes, ed. John Stachel et al., accompanied by English translation by Anna Beck, with Peter Havas, consultant (Princeton, NJ: Princeton University Press, 1987–).

AE, *Peace*   Albert Einstein, *Einstein on Peace*, ed. Otto Nathan and Heinz Norden (New York: Schocken Books, 1960).

AP   Abraham Pais, *"Subtle is the Lord . . .": The Science and the Life of Albert Einstein* (New York: Oxford University Press, 1982).

HE   Gerald Holton and Yehuda Elkana, eds., *Albert Einstein: Historical and Cultural Perspectives: The Centennial Symposium in Jerusalem* (Princeton, NJ: Princeton University Press, 1982).

HSPS   *Historical Studies in the Physical and Biological Sciences*

JM   Christa Jungnickel and Russell McCormmach, *The Intellectual Mastery of Nature: Theoretical Physics from Ohm to Einstein*, 2 vols. (Chicago: University of Chicago Press, 1986)

JS   *Einstein's Miraculous Year: Five Papers That Changed the Face of Physics*, ed. John Stachel, assisted by Trevor Lipscombe, Alice Calaprice, and Sam Elworthy (Princeton, NJ: Princeton University Press, 1998).

LP   Lewis Pyenson, *The Young Einstein: The Advent of Relativity* (Boston: Adam Hilger, 1985).

CHAPTER I. THE EINSTEIN AGE

1. John Herman Randall, *The Making of the Modern Mind: A Survey of the Intellectual Background of the Present Age* (New York: Columbia University Press, 1954/1976), pp. 253–281.

2. Frederic Golden, "Albert Einstein (1879–1955): Person of the Century," *Time*, December 31, 1999, p. 62.

3. Thomas S. Kuhn, *The Structure of Scientific Revolutions*, 2nd ed. (Chicago: University of Chicago Press, 1970).

4. For Einstein, two works are available: Dennis P. Ryan, ed., *Einstein and the Humanities* (New York: Greenwood Press, 1987); and Alan J. Friedman and Carol C. Donley, *Einstein as Myth and Muse* (Cambridge: Cambridge University Press, 1985).

5. JM.

CHAPTER 2. DÉBUT DE SIÈCLE

1. Albrecht Fölsing, *Albert Einstein: A Biography*, trans. Ewald Osers (New York: Viking, 1997), p. 11; and "A. Einstein: Image and Impact," American Institute of Physics, http://www.aip.org/history/einstein (accessed March 29, 2004).

2. See, for instance, LP, pp. 35–57.

3. Max Jammer, *Einstein and Religion: Physics and Theology* (Princeton, NJ: Princeton University Press, 1999), p. 18.

4. AE, *Notes*, p. 3.

5. Ibid., p. 5.

6. Isaiah Berlin, "Einstein and Israel," in HE, p. 289. See also Jammer, *Einstein and Religion*, pp. 43–48.

7. AE, *Papers*, 1:239–41, 277–78.

8. Einstein-Marić correspondence, ibid.

9. Ibid., pp. 60–62.

10. Einstein, letter to Marić, October 3, 1900; Marić to Einstein, ca. 1900; and Einstein to Marić, August 1, 1900, all in Albert Einstein and Mileva Marić, *The Love Letters*, ed. Jürgen Renn and Robert Schulmann, trans. Shawn Smith (Princeton, NJ: Princeton University Press, 1992), pp. 36, 18, 21. Among the many "love exposés" are Dennis Overbye, *Einstein in Love: A Scientific Romance* (New York: Viking Press, 2000); and Andrea Gabor, "Mileva Maric Einstein," in Gabor, *Einstein's Wife: Work and Marriage in the Lives of Five Great Twentieth-Century Women* (New York: Viking, 1995), pp. 3–32.

11. John Herman Randall, "The Newtonian World-Machine," chap. 11 in *The Making of the Modern Mind: A Survey of the Intellectual Background of the Present Age* (New York: Columbia University Press, 1954/1976).

12. Martin J. Klein, "Mechanical Explanation at the End of the Nineteenth Century," *Centaurus* 17 (1972): 58–82.

13. Betty Jo Teeter Dobbs and Margaret C. Jacob, *Newton and the Culture of Newtonianism* (Amherst, NY: Humanity Books, 1994).

14. Edmund Whittaker, *A History of the Theories of Aether and Electricity*, vol. 1 (London: T. Nelson, 1910; reprint, New York: Tomash Publishers and American Institute of Physics, 1987).

15. AE, *Notes*, p. 25; and Albert Einstein and Leopold Infeld, *The Evolution of Physics: The Growth of Ideas from Early Concepts to Relativity and Quanta* (New York: Simon and Schuster, 1938).

16. AE, *Papers*, 1:6–9.

17. Ibid., 2:3–40.

18. AE, *Ideas*, p. 324.

19. Russell McCormmach, "Einstein, Lorentz, and the Electron Theory," *HSPS* 2 (1970): 41–87. The papers of 1905 are available in a new English translation with editorial annotation, in JS.

20. John Theodore Merz, *A History of European Thought in the Nineteenth*

Century, 4 vols. (London: William Blackwood and Sons, 1904–1912; reprint, New York: Dover, 1965).

21. Ibid., and John Blackmore, ed., *Ernst Mach—a Deeper Look: Documents and New Perspectives* (Boston: Kluwer, 1992).

22. Fritz K. Ringer, *The Decline of the German Mandarins: The German Academic Community, 1890–1933* (Cambridge, MA: Harvard University Press, 1969).

23. AE, *Ideas*, pp. 224–27.

24. Ibid., pp. 220–23.

25. AE, *Papers*, 1:28; see also Russell McCormmach, editor's foreword to *HSPS* 7 (1976): xi–xxxv.

26. Paul Forman, John L. Heilbron, and Spencer Weart, "Physics circa 1900: Personnel, Funding, and Productivity of the Academic Establishments," *HSPS* 5 (1975): 31.

27. Theodore Lyman, quoted by Edwin C. Kemble, interviews by T. S. Kuhn and J. H. Van Vleck, May 11, 1962–October 2, 1963 (Niels Bohr Library, American Institute of Physics, College Park, MD).

28. JM, 2:160.

29. Ibid., p. 287.

30. AE, *Notes*, pp. 17, 5.

31. Quoted by Stanley Goldberg, "Max Planck's Philosophy of Nature and His Elaboration of the Special Theory of Relativity," *HSPS* 7 (1976): 142.

CHAPTER 3. TECHNICAL EXPERT, THIRD CLASS

1. John Theodore Merz, *A History of European Thought in the Nineteenth Century* (London: William Blackwood and Sons, 1904–12; reprint, New York: Dover, 1965), 3:391; JM, 2:217–20.

2. Martin J. Klein, "Fluctuations and Statistical Physics in Einstein's Early Work," in HE, pp. 39–58. Einstein's early papers may be found in AE, *Papers*, 2:41–55.

3. English translations of Einstein's dissertation and the 1905 paper are provided in JS, pp. 45–69, 85–98. These and related papers are also translated in Albert Einstein, *Investigations on the Theory of Brownian Movement*, ed. R. Fürth, trans. A. D. Cowper (New York: Dover, 1926/1956), as well in the English edition of AE, *Papers*.

4. AE, *Papers*, 2:107.

5. Klein, "Fluctuations and Statistical Physics"; and Seiya Abiko, "On the Chemico-Thermal Origins of Special Relativity," *HSPS* 22 (1991): 1–24.

6. Russell McCormmach, "H. A. Lorentz and the Electromagnetic View of Nature," *Isis* 61 (1970): 459–97; Arthur I. Miller, *Albert Einstein's Special Theory of Relativity: Emergence (1905) and Early Interpretation (1905–1911)* (Reading, MA: Addison-Wesley, 1981), pp. 45–54.

7. AE, *Papers*, 1:316.

8. The letters may be found in AE, *Papers*, vols. 1 and 5, and in English translation in Albert Einstein and Mileva Marić, *The Love Letters*, ed. Jürgen Renn and Robert Schulmann, trans. Shawn Smith (Princeton, NJ: Princeton University Press, 1992).

9. Einstein and Marić, *The Love Letters*, pp. 39, 72–73.

10. AE, *Papers*, 1:xxxix; Einstein and Marić, *The Love Letters*, pp. xix–xxi.

11. JS, p. 159.

12. John Stachel, *Einstein from 'B' to 'Z'* (Boston: Birkhäuser, 2001).

13. Andrea Gabor, "Mileva Maric Einstein," in Gabor, *Einstein's Wife: Work and Marriage in the Lives of Five Great Twentieth-Century Women* (New York: Viking, 1995), pp. 3–32.

14. Miller, *Albert Einstein's Special Theory of Relativity*; AP.

15. Miller, *Albert Einstein's Special Theory of Relativity*, pp. 29–32; Stanley Goldberg, "The Lorentz Theory of Electrons and Einstein's Theory of Relativity," *American Journal of Physics* 37 (1969): 982–94.

16. Miller, *Albert Einstein's Special Theory of Relativity*; Goldberg, "The Lorentz Theory of Electrons"; Tetu Hirosige, "The Ether Problem, the Mechanistic Worldview, and the Origins of the Theory of Relativity," *HSPS* 7 (1976): 3–82.

17. Goldberg, "The Lorentz Theory of Electrons"; Stanley Goldberg, "Henri Poincaré and Einstein's Theory of Relativity," *American Journal of Physics* 35 (1967): 934–44; Miller, *Albert Einstein's Special Theory of Relativity*, pp. 40–41; Arthur I. Miller, "A Study of Henri Poincaré's 'Sur la Dynamique de l'Electron,'" *Archives for History of Exact Sciences* 10 (1973): 207–328.

18. AE, *Papers*, 2:274–310; JS, pp. 123–59.

19. Nontechnical explanations of this theory are listed in the bibliography, as are some of the many studies of its history.

20. Edmund Whittaker, *A History of the Theories of Aether and Electricity*, vol. 2 (New York: Philosophical Library, 1954; New York: Tomash Publishers and American Institute of Physics, 1987), pp. 27–77.

21. AE, *Ideas*, pp. 227–32; Russell McCormmach, "Einstein, Lorentz, and the Electron Theory," *HSPS* 2 (1970): 41–87.

22. Gerald Holton, "Einstein, Michelson, and the 'Crucial' Experiment," in *Thematic Origins of Scientific Thought: Kepler to Einstein*, rev. ed. (Cambridge, MA: Harvard University Press, 1988).

23. JS, pp. 37–38.

24. Goldberg, "The Lorentz Theory of Electrons."

25. AE, *Papers*, 2:312–14; JS, pp. 161–64.

26. AE, *Papers*, 1:149–69; JS, pp. 177–98.

27. Thomas S. Kuhn, *Black-Body Theory and the Quantum Discontinuity, 1894–1912* (New York: Oxford University Press, 1978).

28. See ibid., pp. 170–82; Martin J. Klein, "Einstein's First Paper on Quanta," *Natural Philosopher* 2 (1963): 59–86; Max Jammer, *The Conceptual Development of Quantum Mechanics* (New York: McGraw-Hill, 1966).

29. JS, p. 177.

30. Kuhn, *Black-Body Theory*, pp. 170–87; Jammer, *The Conceptual Development of Quantum Mechanics*, pp. 26–46; JS, introduction, pp. 167–76.

31. AE, *Papers*, 2:134–48; Martin J. Klein, "Einstein, Specific Heats, and the Early Quantum Theory," *Science* 148 (1965): 173–80.

32. AE, *Papers*, 2:xxv, 74, 147.

33. Ibid., 2:581–82; McCormmach, "Einstein, Lorentz, and the Electron Theory; Martin J. Klein, "Einstein and the Wave-Particle Duality," *Natural Philosopher* 3 (1964): 1–49.

CHAPTER 4. IMPACT ON PHYSICS: RELATIVITY

1. Stanley Goldberg, *Understanding Relativity: Origin and Impact of a Scientific Revolution* (Cambridge, MA: Birkhäuser, 1984). See also Thomas F. Glick, ed., *The Comparative Reception of Relativity* (Boston: D. Reidel, 1987); Loren R. Graham, "The Reception of Einstein's Ideas: Two Examples from Contrasting Political Cultures," in HE, pp. 107–36; and Jean Eisenstaedt and A. J. Kox, eds., *Studies in the History of General Relativity* (Boston: Birkhäuser, 1992).

2. LP, pp. 80–100.

3. Quoted by Stanley Goldberg, "The Lorentz Theory of Electrons and Einstein's Theory of Relativity," *American Journal of Physics* 37 (1969): 991.

4. Albert Einstein and Arnold Sommerfeld, *Briefwechsel*, ed. Armin Hermann (Basel: Schwabe, 1968), p. 20.

5. JM, 2:240; Stanley Goldberg, "Max Planck's Philosophy of Nature and His Elaboration of the Special Theory of Relativity," *HSPS* 7 (1976): 125–60.

6. LP, pp. 194–214.

7. Quoted by Goldberg, "Max Planck's Philosophy of Nature," p. 157.

8. JM, 2:294–99.

9. AE, *Ideas*, pp. 220–21.

10. Mileva Marić, letter to Helene Savić, March 12, 1913, quoted in Desanka Trbuhović-Gjurić, *Im Schatten Albert Einsteins: Das tragische Leben der Mileva Einstein-Marić* (Bern: Verlag Paul Haupt, 1983), p. 104.

11. AE, *Papers*, vol. 5.

12. AE, *Ideas*, p. 9.

13. LP, pp. 137–57; JM, 2:247.

14. LP, pp. 101–36.

15. AE, *Ideas*, pp. 224–27.

16. Russell McCormmach, editor's foreword to *HSPS* 7 (1976): xi–xxxv.

17. AE, *Papers*, 2:503–507; LP, p. 223.

18. LP, pp. 151–54; McCormmach, editor's foreword to *HSPS* 7, p. xxi.

19. AE, *Papers*, 2:148.

20. JM, 2:332; Russell McCormmach, "Einstein, Lorentz, and the Electron Theory," *HSPS* 2 (1970): 41–87.

21. The history is surveyed by, among others, AP, pp. 177–296; and Albert Einstein and Leopold Infeld, *The Evolution of Physics: The Growth of Ideas from Early Concepts to Relativity and Quanta* (New York: Simon and Schuster, 1938).

22. Einstein and Infeld, *The Evolution of Physics*, p. 233.

23. AE, *Papers*, 5:505.

24. Albert Einstein, letter to Paul Hertz, August 22, 1915; AE, *Papers*, 8A:163.

25. For technical details, see AP, pp. 281–88.

26. For more on this subject, see, among many others, Stephen W. Hawking, *The Theory of Everything: The Origin and Fate of the Universe* (New York: New Millennium, 2003) and *A Brief History of Time: The Updated and Expanded Tenth Anniversary Edition* (New York: Bantam, 1998).

27. For more on this subject, see Hawking, *The Theory of Everything*, and on Oppenheimer, see D. Cassidy, *J. Robert Oppenheimer and the American Century* (New York: Pi Press, 2004).

CHAPTER 5. IMPACT ON PHYSICS: THE QUANTUM

1. AE, *Papers*, 2:234.

2. AP, pp. 357–469. See also Max Jammer, *The Conceptual Development of Quantum Mechanics* (New York: McGraw-Hill, 1966). An exhaustive technical history of quantum theory is provided by J. Mehra and H. Rechenberg, *The Historical Development of Quantum Mechanics*, 6 vols. (New York: Springer-Verlag, 1982–2001).

3. Planck et al., June 12, 1913, AE, *Papers*, 5:527.

4. Quoted in AP, p. 357.

5. AE, *Papers*, 2:564–65; Martin J. Klein, "Einstein and the Wave-Particle Duality," *Natural Philosopher* 3 (1964): 1–49; Thomas S. Kuhn, *Black-Body Theory and the Quantum Discontinuity, 1894–1912* (New York: Oxford University Press, 1978).

6. AE, *Papers*, 2:378–91; Martin J. Klein, "Einstein, Specific Heats, and the Early Quantum Theory," *Science* 148 (1965): 173–80.

7. JM, 2:321.

8. See D. Cassidy, *Uncertainty: The Life and Science of Werner Heisenberg* (New York: W. H. Freeman, 1992).

9. Paul Forman, "The Financial Support and Political Alignment of Physicists in Weimar Germany," *Minerva* 12 (1974): 49–66.

10. Karl T. Compton, *Nature* 139 (1937): 238–39.

11. See, for instance, Arthur I. Miller, *Imagery in Scientific Thought: Creating Twentieth-Century Physics* (Boston: Birkhäuser, 1984).

12. Quoted by Heisenberg, *Physics and Beyond* (New York: Harper and Row, 1971), p. 63.

13. Erwin Schrödinger, "Über das Verhältnis der Heisenberg-Born-Jordanschen Quantenmechanik zu der Meinen," *Annalen der Physik* 79 (1926): 734; Albert Einstein, letter to Schrödinger, April 26, 1926, in *Briefe zur Wellenmechanik*, ed. K. Przibram (Vienna: Springer-Verlag, 1963), p. 26.

14. The philosophical implications have been explored widely in the literature, for instance Ernst Cassirer, *Determinism and Indeterminism in Modern Physics*, trans. Theodor Benfey (New Haven, CT: Yale University Press, 1956).

15. Quoted in Cassidy, *Uncertainty*, p. 250.

16. AP, pp. 442–57; Martin J. Klein, "Einstein and the Wave-Particle Duality," *Natural Philosopher* 3 (1964): 1–49; Niels Bohr, "Discussion with Einstein on Epistemological Problems in Atomic Physics," in *Albert Ein-*

stein: Philosopher-Scientist, ed. Paul Arthur Schilpp (LaSalle, IL: Open Court, 1949/1969), 1:199–241.

17. Albert Einstein, Hedwig Born, and Max Born, Briefwechsel, 1916–1955, ed. Max Born (Munich: Nymphenburger Verlagshandlung, 1969), pp. 129–30.

18. See, for instance, Katherine R. Sopka, Quantum Physics in America, 1920–1935 (New York: Arno Press, 1980).

19. Jamie Sayen, Einstein in America: The Scientist's Conscience in the Age of Hitler and Hiroshima (New York: Crown, 1985).

20. JM, 2:372.

CHAPTER 6. RELATIVITY REACHES THE PUBLIC

1. Quoted in Loren R. Graham, "The Reception of Einstein's Ideas: Two Examples from Contrasting Political Cultures," in HE, p. 107.

2. Gerald Holton, "Introduction: Einstein and the Shaping of Our Imagination," in HE, pp. vii–xxxii; reprint, Holton, Physics, History, and Other Passions (New York: American Institute of Physics, 1994).

3. Holton, "Introduction," pp. xxii, xxvi; see also Alan J. Friedman and Carol C. Donley, Einstein as Myth and Muse (Cambridge: Cambridge University Press, 1985), p. 20.

4. Paul Forman, "Weimar Culture, Causality, and Quantum Theory, 1918–1927," HSPS 3 (1971): 1–115; Roland N. Stromberg, An Intellectual History of Modern Europe, 2nd ed. (Englewood Cliffs, NJ: Prentice-Hall, 1975), p. 377.

5. Quoted in Allie Vibert Douglas, The Life of Arthur Stanley Eddington (London: Thomas Nelson and Sons, 1956), p. 43.

6. Albert Einstein, Hedwig Born, and Max Born, Briefwechsel, 1916–1955, ed. Max Born (Munich: Nymphenburger Verlagshandlung, 1969), p. 59.

7. Friedman and Donley, Einstein as Myth and Muse, p. 20.

8. Forman, "Weimar Culture."

9. Armin Hermann, The New Physics: The Route into the Atomic Age, trans. D. Cassidy (Bonn and Bad Godesberg: InterNationes, 1979), p. 54.

10. Fritz Stern, Einstein's German World (Princeton, NJ: Princeton University Press, 1999), chap. 7; Thomas Levenson, Einstein in Berlin (New York: Bantam Books, 2003); Albrecht Fölsing, Albert Einstein: A Biography, trans. Ewald Osers (New York: Viking, 1997), chap. 23.

11. AE, *Peace*, p. 2.

12. Fölsing, *Albert Einstein*, pp. 465–68.

13. Alan D. Beyerchen, *Scientists under Hitler: Politics and the Physics Community in the Third Reich* (New Haven, CT: Yale University Press, 1977), pp. 79–122.

14. Ibid., pp. 111–15; Paul Forman, "The Financial Support and Political Alignment of Physicists in Weimar Germany," *Minerva* 12 (1974): 49–66.

15. Quoted in LP, p. 232.

16. Quoted in Beyerchen, *Scientists under Hitler*, p. 92.

17. Ibid., p. 96.

18. Andreas Kleinert, "Lenard, Stark und die Kaiser-Wilhelm-Gesellschaft," *Physikalische Blätter* 36 (1980): 35.

19. Beyerchen, *Scientists under Hitler*, pp. 123–40.

20. AE, *Peace*, pp. 222–23.

21. Philipp Frank, *Albert Einstein: His Life and Times* (New York: A. A. Knopf, 1947), pp. 416–18.

22. Graham, "The Reception of Einstein's Ideas"; Loren R. Graham, *Science, Philosophy, and Human Behavior in the Soviet Union* (New York: Columbia University Press, 1987).

23. Eddington, quoted in Graham, "The Reception of Einstein's Ideas," p. 113; James Jeans, *Physics and Philosophy* (New York: Macmillan, 1946), p. 307.

24. Jeans, *Physics and Philosophy*, p. 216.

25. Graham, "The Reception of Einstein's Ideas"; Loren R. Graham, *Science in Russia and the Soviet Union: A Short History* (New York: Cambridge University Press, 1993).

26. Edward Friedman, "Political Origins and Significance of China's Einstein Centennial," in *Einstein and the Humanities*, ed. Dennis P. Ryan (New York: Greenwood Press, 1987), p. 154.

CHAPTER 7. CULTURAL RESONANCES

1. Both quotes in Fritz Stern, "Einstein's Germany," in HE, p. 320.

2. Andrew Paul Ushenko, "Einstein's Influence on Contemporary Philosophy," in *Albert Einstein: Philosopher-Scientist*, ed. Paul Arthur Schilpp (LaSalle, IL: Open Court, 1949/1969), 2:608; Peter Galison, *Einstein's Clocks, Poincaré's Maps: Empires of Time* (New York: W. W. Norton, 2003).

3. Quoted by Ushenko, "Einstein's Influence on Contemporary Philosophy," p. 628.

4. A classic history is Viktor Kraft, *The Vienna Circle: The Origin of Neo-Positivism, a Chapter in the History of Recent Philosophy*, trans. Arthur Pap (New York: Philosophical Library, 1953); translations of some original papers are included in *Logical Positivism*, ed. A. J. Ayer (New York: Free Press, 1959).

5. Moritz Schlick, "The Turning Point in Philosophy," in Ayer, ed., *Logical Positivism*, p. 57, his italics.

6. Rudolf Carnap, "The Elimination of Metaphysics through Logical Analysis of Language," in Ayer, ed., *Logical Positivism*, pp. 76–77, his italics.

7. Thomas S. Kuhn, *The Structure of Scientific Revolutions*, 2nd ed. (Chicago: University of Chicago Press, 1970).

8. Some early responses appeared in *Criticism and the Growth of Knowledge*, ed. Imre Lakatos and Alan Musgrave (Cambridge: Cambridge University Press, 1972); more recently, Paul Hoyningen-Huene, *Recasting Scientific Revolutions: Thomas S. Kuhn's Philosophy of Science*, trans. Alexander T. Levine (Chicago: University of Chicago Press, 1993); and the sociological critique by Steve Fuller, *Thomas Kuhn: A Philosophy for Our Times* (Chicago: University of Chicago Press, 2000).

9. Tristan Tzara, "Dada Manifesto," quoted in *Paths to the Present: Aspects of European Thought from Romanticism to Existentialism*, ed. Eugen Weber (New York: Dodd, Mead, 1960), p. 251.

10. Quoted in Gerald Holton, "Einstein's Influence on the Culture of Our Time," in Holton, *Einstein, History, and Other Passions* (Reading, MA: Addison-Wesley, 1996).

11. See Arthur I. Miller, *Einstein, Picasso: Space, Time, and the Beauty that Causes Havoc* (New York: Basic Books, 2001); Alan J. Friedman and Carol C. Donley, *Einstein as Myth and Muse* (Cambridge: Cambridge University Press, 1985), pp. 21–25.

12. Jack Burnham, *Beyond Modern Sculpture: The Effects of Science and Technology on the Sculpture of This Century* (New York: G. Braziller, 1968), p. 256.

13. Quoted in Weber, ed., *Paths to the Present*, p. 237.

14. Horst de la Croix and Richard G. Tansey, *Gardner's Art through the Ages*, 7th ed. (New York: Harcourt Brace Jovanovich, 1980), p. 839.

15. Ibid., p. 842; see also Sigfried Giedion, *Space, Time, and Architecture: The Growth of a New Tradition*, 4th ed. (Cambridge, MA: Harvard University Press, 1965).

16. Burnham, *Beyond Modern Sculpture*, pp. 254–62.

17. Holton, "Introduction," pp. xvii–xx; see also Friedman and Donley, *Einstein as Myth and Muse*, pp. 83–109.

18. Quoted in D. Cassidy, *Uncertainty: The Life and Science of Werner Heisenberg* (New York: W. H. Freeman, 1992), p. 40.

19. Carol Donley, "'Springtime of the Mind': Poetic Responses to Einstein and Relativity," in *Einstein and the Humanities*, ed. Dennis P. Ryan (New York: Greenwood Press, 1987), pp. 119–24.

20. e. e. cummings, *Complete Poems, 1904–1962*, rev. ed., ed. George J. Firmage (New York: Liveright, 1991), p. 317.

21. William Carlos Williams, *Selected Essays* (New York: Random House, 1954), pp. 238, 281.

22. Hyatt Howe Waggoner, *The Heel of Elohim: Science and Values in Modern American Poetry* (Norman, OK: University of Oklahoma Press, 1950), pp. 143–44.

23. Archibald MacLeish, *Collected Poems, 1917–1982* (Boston: Houghton Mifflin, 1985), p. 144.

CHAPTER 8. SCIENCE AND CONSCIENCE

1. AP, pp. 473–78; Bernard T. Feld, "Einstein and the Politics of Nuclear Weapons," in HE, pp. 369–93; Alice Kimball Smith, *A Peril and a Hope: The Scientists' Movement in America, 1945–47* (Cambridge, MA: MIT Press, 1971).

2. AE, *Peace*, p. 218.

3. Isaiah Berlin, "Einstein and Israel," in HE, pp. 281–92.

4. AE, *Ideas*, p. 171.

5. Quoted in Berlin, "Einstein and Israel," p. 285.

6. Ibid., p. 288.

7. AE, *Ideas*, p. 180.

8. Ibid., p. 184.

9. Ibid., p. 173.

10. Ibid., p. 201.

11. AE, *Peace*, p. 573.

12. Ibid., pp. 294–96. For the early history of nuclear weapons and the Manhattan Project, see Richard Rhodes, *The Making of the Atomic Bomb* (New York: Simon and Schuster, 1986).

13. Quoted in Albrecht Fölsing, *Albert Einstein: A Biography*, trans. Ewald Osers (New York: Viking, 1997), p. 725.

14. Quoted in Feld, "Einstein and the Politics of Nuclear Weapons," p. 376; Einstein's emphasis.

15. AE, *Peace*, p. 125.

16. Ibid., pp. 229, 241.

17. The background to the refused security clearance is discussed by Fred Jerome, *The Einstein File: J. Edgar Hoover's Secret War against the World's Most Famous Scientist* (New York: St. Martin's Press, 2002).

18. AE, *Peace*, pp. 304–305; Rhodes, *The Making of the Atomic Bomb*, pp. 635–39; Smith, *A Peril and a Hope*, pp. 27–28.

19. AE, *Peace*, p. 555.

20. Feld, "Einstein and the Politics of Nuclear Weapons," p. 382; AE, *Peace*, pp. 410–12.

21. These developments are further discussed by Richard Rhodes, *Dark Sun: The Making of the Hydrogen Bomb* (New York: Simon and Schuster, 1995).

22. Majority and minority statements, General Advisory Committee, Atomic Energy Commission, October 30, 1949, reprinted in *The American Atom: A Documentary History of Nuclear Policies from the Discovery of Fission to the Present, 1939–1984*, ed. Robert C. Williams and Philip L. Cantelon (Philadelphia: University of Pennsylvania Press, 1984), pp. 126–27.

23. Einstein, "Arms Can Bring No Security," *Bulletin of the Atomic Scientists* 6 (March 1950): 71; revised version in AE, *Peace*, pp. 520–22.

24. AE, *Peace*, p. 534.

25. The impact on science has been explored, for instance, by Jessica Wang, *American Science in an Age of Anxiety: Scientists, Anticommunism, and the Cold War* (Chapel Hill: University of North Carolina Press, 1999).

26. Jerome, *The Einstein File*, analyzes the file.

27. AE, *Peace*, pp. 546–47. The letter appeared in the *New York Times*, June 12, 1953.

28. AE, *Peace*, p. 635.

SELECT
BIBLIOGRAPHY

The literature on Einstein and his influence is vast and still growing. The items listed below include only a sampling of the technical and nontechnical books for general audiences. Additional references and suggestions for further reading may be found in the notes and in the bibliography of the Web page "A. Einstein: Image and Impact," http://www.aip.org/history/einstein, sponsored by the Center for History of Physics, American Institute of Physics.

Bernstein, Jeremy. *Albert Einstein and the Frontiers of Physics*. New York: Oxford University Press, 1996.

Bodanis, David. $E = mc^2$: *A Biography of the World's Most Famous Equation*. New York: Walker, 2000.

Brian, Denis. *Einstein: A Life*. New York: John Wiley, 1996.

Calaprice, Alice, and Freeman J. Dyson, eds. *The Expanded Quotable Einstein*. Princeton, NJ: Princeton University Press, 2000.

Einstein, Albert. *Autobiographical Notes: A Centennial Edition*. Edited by Paul Arthur Schilpp. LaSalle, IL: Open Court, 1991.

———. *The Collected Papers of Albert Einstein.* Multiple volumes. Edited by John Stachel et al. Princeton, NJ: Princeton University Press, 1987–.

———. *The Collected Papers of Albert Einstein* [English translation]. Multiple volumes. Translated by Anna Beck, with Peter Havas, consultant. Princeton, NJ: Princeton University Press, 1987–.

———. *Einstein on Peace.* Edited and translated by Otto Nathan and Heinz Norden. New York: Schocken Books, 1960.

———. *Ideas and Opinions.* Edited by Carl Seelig. Translated by Sonja Bargmann. 1952; reprint, New York: Modern Library, 1994.

———. *Out of My Later Years.* 1950; reprint, New York: Wings Books of Random House, 1993.

———. *Relativity: The Special and the General Theory.* Translated by Robert W. Lawson. 1916; reprint, New York: Crown, 1995.

Einstein, Albert, and Leopold Infeld. *The Evolution of Physics: The Growth of Ideas from Early Concepts to Relativity and Quanta.* New York: Simon and Schuster, 1938.

Einstein, Albert, and Mileva Marić. *The Love Letters.* Edited by Jürgen Renn and Robert Schulmann. Translated by Shawn Smith. Princeton, NJ: Princeton University Press, 1992.

Fölsing, Albrecht. *Albert Einstein: A Biography.* Translated by Ewald Osers. New York: Viking, 1997.

Frank, Philipp. *Albert Einstein: His Life and Times.* New York: A. A. Knopf, 1947.

French, A. P., ed. *Einstein: A Centenary Volume.* Cambridge, MA: Harvard University Press, 1979.

Friedman, Alan J., and Carol C. Donley. *Einstein as Myth and Muse.* Cambridge: Cambridge University Press, 1985.

Galison, Peter. *Einstein's Clocks, Poincaré's Maps: Empires of Time.* New York: W. W. Norton, 2003.

Gardner, Martin. *The Relativity Explosion.* 1982; reprint, New York: Vintage Books, 1976.

Glick, Thomas F., ed. *The Comparative Reception of Relativity.* Boston: D. Reidel, 1987.

Goldberg, Stanley. *Understanding Relativity: Origin and Impact of a Scientific Revolution.* Cambridge, MA: Birkhäuser, 1984.

Holton, Gerald. *Thematic Origins of Scientific Thought: Kepler to Einstein.* Rev. ed. Cambridge, MA: Harvard University Press, 1988.

———. *Einstein, History, and Other Passions.* Cambridge, MA: Harvard University Press, 2000.

Holton, Gerald, and Yehuda Elkana, eds. *Albert Einstein: Historical and Cultural Perspectives. The Centennial Symposium in Jerusalem.* Princeton, NJ: Princeton University Press, 1982.

Jammer, Max. *The Conceptual Development of Quantum Mechanics.* New York: McGraw-Hill, 1966.

——. *Einstein and Religion: Physics and Theology.* Princeton: Princeton University Press, 1999.

Jerome, Fred. *The Einstein File: J. Edgar Hoover's Secret War against the World's Most Famous Scientist.* New York: St. Martin's, 2002.

Jungnickel, Christa, and Russell McCormmach. *The Intellectual Mastery of Nature: Theoretical Physics from Ohm to Einstein.* 2 vols. Chicago: University of Chicago Press, 1986.

Kerszberg, Pierre. *The Invented Universe: The Einstein–De Sitter Controversy (1916–1917) and the Rise of Relativistic Cosmology.* New York: Oxford University Press, 1989.

Kuhn, Thomas S. *Black-Body Theory and the Quantum Discontinuity, 1894–1912.* New York: Oxford University Press, 1978.

Levenson, Thomas. *Einstein in Berlin.* New York: Bantam Books, 2003.

Miller, Arthur I. *Albert Einstein's Special Theory of Relativity: Emergence (1905) and Early Interpretation (1905–1911).* Reading, MA: Addison-Wesley, 1981.

——. *Einstein, Picasso: Space, Time, and the Beauty that Causes Havoc.* New York: Basic Books, 2001.

Pais, Abraham. *Einstein Lived Here: Essays for the Layman.* New York: Oxford University Press, 1994.

——. *"Subtle is the Lord . . .": The Science and the Life of Albert Einstein.* New York: Oxford University Press, 1982.

Parker, Barry. *Einstein: The Passions of a Scientist.* Amherst, NY: Prometheus Books, 2003.

Pyenson, Lewis. *The Young Einstein: The Advent of Relativity.* Boston: Adam Hilger, 1985.

Russell, Bertrand. *The ABC of Relativity.* 3rd, rev. ed. London: George Allen and Unwin, 1958.

Ryan, Dennis P., ed. *Einstein and the Humanities.* New York: Greenwood Press, 1987.

Sayen, Jamie. *Einstein in America: The Scientist's Conscience in the Age of Hitler and Hiroshima.* New York: Crown, 1985.

Schilpp, Paul Arthur, ed. *Albert Einstein: Philosopher-Scientist.* 2 vols. 1949; reprint, LaSalle, IL: Open Court, 1969.

Stachel, John. *Einstein from 'B' to 'Z'*. Boston: Birkhäuser, 2001.

———, ed. *Einstein's Miraculous Year: Five Papers That Changed the Face of Physics*. Assisted by Trevor Lipscombe, Alice Calaprice, and Sam Elworthy. Princeton, NJ: Princeton University Press, 1998.

Stern, Fritz. *Einstein's German World*. Princeton, NJ: Princeton University Press, 2001.

Williams, L. Pearce, ed. *Relativity Theory: Its Origin and Impact on Modern Thought*. New York: Wiley, 1968.

Woolf, Harry, ed. *Some Strangeness in the Proportion: A Centennial Symposium to Celebrate the Achievements of Albert Einstein*. Reading, MA: Addison-Wesley, 1980.

INDEX